the Octopus's Garden

the Octopus's Garden

HYDROTHERMAL VENTS AND OTHER MYSTERIES OF THE DEEP SEA

Cindy Lee Van Dover

Illustrations by Karen Jacobsen

ADDISON-WESLEY PUBLISHING COMPANY
*Reading, Massachusetts   Menlo Park, California   New York
Don Mills, Ontario   Wokingham, England   Amsterdam   Bonn
Sydney   Singapore   Tokyo   Madrid   San Juan
Paris   Seoul   Milan   Mexico City   Taipei*

Library of Congress Cataloging-in-Publication Data

Van Dover, Cindy.
    The octopus's garden : hydrothermal vents and other mysteries
of the deep sea / Cindy Lee Van Dover.
        p.   cm.
    Includes index.
    ISBN 0-201-40770-1
    1. Deep-sea ecology.   2. Hydrothermal vent fauna.   3. Alvin
(Submarine)   I. Title.
QH541.5.D35V35   1996
574.5′2636—dc20   95–21382
                                                            CIP

Grateful acknowledgment is made to
Startling Music, Ltd., for the excerpt from *The Octopus's
Garden.* Copyright © 1969, Startling Music, Ltd.
Portions of this book have appeared, in somewhat different form,
in *Smithsonian, Discover,* and *Oceanus,* and are reprinted here
with their permission.

Jacket design by Lynne Reed
Text illustrations by Karen Jacobsen
Text design by Karen Savary
Set in 11-point Horley Old Style by Carol Woolverton Studio

1 2 3 4 5 6 7 8 9 10-DOH-9998979695
First printing, December 1995

For
The Alvin *Group*
and the
Crew of the R/V Atlantis II

I'd like to be
Under the Sea
In an Octopus's Garden
In the Shade

Contents

Acknowledgments   *xiii*

Introduction   *1*

1. *Beyond the Edge*   *9*

2. *The Right Stuff*   *19*

3. Alvin   *29*

4. *The Abyssal Wilderness*   *47*

5. *A Chorus of Tubeworms*   *73*

6. *On Broken Spur*   *89*

7. *Black Smokers*   *101*

8. *Chasing Eruptions*   *115*

9. *"Blind" Shrimp*   *127*

10. *Biological Particulars*   *139*

11. *The Trilobite Factor*   *151*

12. *Sediment Plains*   *159*

Epilogue   *169*

Index   *173*

Acknowledgments

M Y LIFE IS RICH WITH MENTORS AND FRIENDS
and colleagues who have helped me find my way
into the Octopus's Garden: Ken Smith, chief scientist of my first oceanographic expedition, and others
whom I first met on that cruise—including Robert
Hessler, Fred Grassle, Howard Sanders, Richard Lutz,
Ruth Turner, Carl Berg—provided the initial entree with
their support and encouragement. Fred Grassle took me
on as a graduate student at Woods Hole, and it was Fred,
Bob Ballard, Geoff Thompson, John Edmond, John
Delaney, and others who took me to sea and taught me
how to use the tools of deep-ocean research. Ralph Hollis
was the first to encourage me to pursue my idea of becoming an *Alvin* pilot. Barrie Walden was bold enough to hire

me as a pilot-in-training, and his words of praise, infre-
quent though they were, helped me to persevere, as did
unfailing support from Craig Dorman, Charlie Hollister,
Dudley Foster, Pat Hickey, Joe Cann, Sarah Little, Carla
Curran, and Laela Sayigh.

My cousin, Gay Van Dover, first started me on a
path toward writing this book with her queries about what
exactly I did for a living and why. Jack Repcheck, then at
Addison-Wesley, sought answers to similar questions.
Cyrisse Jaffee helped me turn a collection of essays into
this book. Karen Jacobsen, illustrator for many scientific
expeditions, joined one of my cruises and brings this
experience to bear in the illustrations that accompany the
text. Along the way, John Murray, Michael Collier, Elaine
Barber, and Keith Markolf read drafts of my prose and
offered kind words. My tenure as McCurdy Scholar at the
Duke University Marine Laboratory in Beaufort, North
Carolina, allowed me to complete the book surrounded by
friends.

It is impossible to express how much I owe to the
Alvin Group and the crew of the *Atlantis II*. One does not
get to the seafloor by oneself, whether as a pilot or as a scientist. The men and women who operate the sub and the
boat are the best in the world and it is always a privilege
to work with them and learn from them.

the Octopus's Garden

Introduction

WHEN I WAS A CHILD, I THOUGHT THAT all the world was known. It was taught to me so. I thought the list of explorers complete, the species and their habitats all cataloged. I thought with envy of sailors who discovered whole new lands. I wished that I had been the first person ever to stand at the edge of the Grand Canyon or beside the hot springs of Yellowstone. I wanted to be the first astronaut to set foot on the moon.

I was twenty-nine years old when I first looked over the rim of the Grand Canyon. Although my visit to the great hole was brief, the experience marked me, settled into my being. It became, I am certain, a subconscious part of what defines me and the way that I look at the world. The aurora borealis was like that, too. I recall exploring the autumnal woods of northern Maine with my backpack and dog, where I chanced upon a fortnight's display of light beams dancing through the midnight skies. I was mesmerized. And I wondered at my naivete; how could I have lived so long in ignorance of the aurora? Like the canyon, the aurora is so magnificent, so awesome, that it seems to be a part of the very soul of nature.

There is another place, a vast expanse of places, really, that rivals the grandeur, mystery, and fantasy of the Grand Canyon or the aurora borealis. You are not likely to go there to see it for yourself, yet I go all the time—to the bottom of the sea, to the Octopus's Garden.

Mine is a scientist's perspective of the deep sea. As a scientist, I am robed with degrees and academic pedigree. I write reams of dry prose with appropriately technical language and what my colleagues consider scientific consequence. But at heart, I confess, I am an amateur naturalist, quick to delight in the unusual nature of a worm, the oddities of a shrimp, the peculiarities of a rock.

By literary standards, the deep sea is a solemn place, mysterious and vaguely sinister, a place of loss, a place unknown, where Tennyson's giant Kraken sleeps below the thunder of the upper deep, where faintest sunlights flee. Only Captain Nemo has voyaged 20,000 leagues through its basins; only the *Nautilus* of fiction has toured the submarine depths of all seven seas. As Monsieur Arronax says, in the last words (words that still apply) of

Jules Verne's tale of Prince Dakkar: "And to that question posed 6,000 years ago in the Book of Ecclesiastes—'Who can fathom the depths of the abyss?'—only two men out of all humanity now have the right to respond. Captain Nemo and I."

There is no real boundary to the part of the planet I think of as the deep sea. Technically, it is defined as oceanic depths greater than a few hundred meters. In my mind, the deep sea encompasses the depths of the open ocean beyond where daylight penetrates—beyond where the sun at noon becomes twilight, beyond darkness, into utter black.

It is said that the seafloor is a desert, a vast and uniform wasteland, all but devoid of life. Textbooks on the shelf in my laboratory say so. But I know that is not true. Life on the seafloor is often unusual and diverse, more the stuff of science fiction; it is occasionally abundant, sometimes spectacularly so. Granted, in many places the diversity and wealth of life are to be appreciated only through a microscope and painstaking identification of small worms and mollusks and crustaceans captured in the soft muds of cores taken from the abyssal plains.

In what has become the Oregon Trail of deep-sea biology, pioneering scientists from the Woods Hole Oceanographic Institution in Massachusetts worked the Gay Head–Bermuda Transect in the 1960s. The transect was a series of oceanographic stations from off the cliffs of Martha's Vineyard, down the shelf and slope of the continental margin, into the deep sea, with a southern terminus at Bermuda. Sampling the seafloor by dredging or coring, the Woods Hole scientists found to their surprise that the biodiversity of the fauna living in the perpetual midnight of deep waters was higher than in shallow regions of the transect. Few of the deepwater species are very familiar, even to specialists. Most abyssal species are lucky if they have names. But the variety of life per acre of abyssal seafloor vies with the richness of life in an equal amount of tropical rain forest.

Deep-sea research was then—and remains—a frontier science. The seafloor is the largest and least known wilderness on our planet. We know more about the surface of Mars and Venus and the back side of the moon than we know about the seafloor. Nearly all of the active extension and subduction of the earth's crust takes place in the submarine environment. It is estimated that more than 20 cubic kilometers of lava—a volume that would cover an area four times the size of Alaska with 1 meter of lava—is produced annually by submarine volcanic systems. More than one million volcanoes of all sizes, from 50-meter relief to full-fledged islands, are predicted to occur in the Pacific Ocean Basin alone. Less than 1 percent of these volcanoes have been charted, a reflection of the embarrassing fact that less than 1 percent of our planet's seafloor has been mapped.

We have reason to think that every process that forced the formation of the planet's surface as we know it today is happening now, somewhere on the seafloor. Most of the annual heat loss from the earth's interior occurs through submarine volcanism. If we are to understand the geological forces that shape our planet, we must understand the geological processes expressed by features on the seafloor.

The potential for fundamental discoveries is high for anyone who goes into the field of deep-sea research with open eyes and an intuitive mind. I sensed all this profoundly when I first began to think about what might lie beneath the surface of the oceans. I also felt the romantic draw of deep-sea research—sailing on the high seas, exotic port stops, a chance to prove myself against the elements. I longed to sail on a working research vessel that would carry me beyond sight of land. There is no lineage of sailors in my ancestry. I am puzzled by this need to sail and can attribute it only to the strong influence of the few seafaring novels that were among the books I loved as a child.

In 1989 I found myself a ship—the *Atlantis II,* mother ship of the research submersible *Alvin.* I had known of *Alvin* for years. I can remember when I first read about her back in the late 1960s in a book on oceanography. My dream of diving in her was born then, even before I was out of grade school, but I had always thought that was about as likely as my going to the moon.

For two and a half years, the *AII* was my only home and with her I sailed in blue water. Not only did I spend

 my years before the mast, I dove beneath the seas; I became a submersible pilot. A handsome pin tucked away in the jewelry box on my dresser is testimony to this accomplishment. The insignia comprises two golden dolphin-fish, one fore and one aft, escorting the *Trieste*, the first manned submersible operated by the United States Navy. Those in the navy would recognize these dolphins as the marque worn by pilots-in-command of deep-diving submersibles and wonder what sweetheart might have given them to me. But I earned my dolphins the day I stood as a civilian in front of a navy board in San Diego, answering a morning's worth of questions about systems and procedures on *Alvin*. Because *Alvin* is certified to dive by the navy, all of her pilots must also be navy certified, although she is operated entirely by civilians working for the Woods Hole Oceanographic Institution. The navy board was the last in a gauntlet of oral exams. With my dolphins, I was given a certificate that announces I am the forty-ninth individual to qualify as a "deep submergence vehicle operator." I became one of the elite, one who had the Right Stuff.

I want to share this privilege I have of diving to the seafloor. There is a beauty to life and landscape there that pleases me. It is life sculpted by extreme and hostile conditions, life that is fragile and all but unknown. It is a world for explorers. I follow modestly in the footsteps of William Beebe, the first scientist to descend the ocean depths in a bathysphere. Beebe wrote of his experiences in a book he called *Half Mile Down*. I echo his words here:

> If one dives and returns to the surface inarticulate with amazement and with a deep realization of the marvel of what he has seen and where he has been, then he deserves to go again and again. If he is unmoved or disappointed, then there remains for him on earth only a longer or shorter period of waiting for death.

1

BEYOND THE EDGE

THE DEEP SEA IS NOT AN OBVIOUS PLACE TO dedicate a life in science. Few of us find our way there. It has none of the enviropolitical cachet of an Amazonian rain forest, Alaskan tundra, or Antarctic ice shelf. When I first became interested in the deep sea, there was not even the fantasia world tenanted by alien-looking and gigantically proportioned tubeworms to attract notice. Their discovery, among many others, was still several years away.

Rachel Carson's world—of shallow waters and inter-tidal zones, the edge of the sea—is the bounded one in which I was complacent for a time, happy to have my environment well defined, circumscribed, and within easy reach. In a small field station on the eastern shore of

Delaware Bay I spent my college summers in the mid-1970s on a tidal clock, working the mudflats at every low tide that coincided with daylight hours, tending to trays of cultured oysters and watching the summer seasons progress through cycles of rapid growth of *Polydora* worms, *Balanus* barnacles, *Molgula* seasquirts, and wild *Crassostrea* oysters that threatened to overwhelm my pedigreed shellfish.

As a naturalist, I kept my eyes open to all the minutiae of life on the exposed flats to the farthest limits of where the tide withdrew. I learned to listen for the tidal bore that sounds like a distant train rumbling past and told me when the water would begin its flood. I could tell by the warmth of the water rising above my shins the very hour that the oysters in my trays would spawn. I watched the moon and and knew as it swelled to full in the warmth of a June night that the horseshoe crabs would gather at the water's edge to mate. I learned to tell the turn of the tide from the tenor of the laughing gulls that fed in raucous rows at the line of dead horseshoe crab eggs piled inches deep, left by the last high tide. I knew where to dig to find the nests of live horseshoe crab eggs, and could tell when the hatch would take place by watching for the moment when I could see, through the clearing egg membranes, the small trilobite larvae spinning and whirling, as if in training for their imminent ride in the wash of the surf.

From this apprenticeship to nature on tidal flats, I

knew the strength of the environment in controlling cycles of marine life. I stumbled upon a paradox when I read of the marine world beyond the edge of my tidal flat, beyond the continental shelf, and into the deep sea where the ordinary progressions of seasons—changes in temperature and sunlight—are absent. What, then, I wondered, controls the tempo of life in the deep sea?

My explorations into deep-sea science were limited at first to the dusty stacks of libraries where I read all I could about the animals that live in the abyss. I discovered that others were asking the same questions I was. But popular support for exploration of the oceans and the seafloor has been burdened by historical misconceptions and ignorance of the deep-sea environment. Properties of the deep ocean—extreme cold, tremendous pressure, darkness—biased scientists of the early 1800s to suggest, perhaps anthropocentrically, that life could not exist below 600 meters. This intuitive but incorrect concept of an "azoic zone" was quickly challenged and thoroughly and repetitively repudiated during a subsequent era of global oceanographic voyages, beginning with the *Challenger* Expedition in the 1870s and ending with the *Galathea* Expedition in the 1950s.

Relying on surface vessels and dredging operations, scientists recovered great quantities of organisms from deep water. But the legacy of the azoic zone, and the idea that the deep sea cannot support life and is therefore dull and uninteresting, still prevail today among laypeople and even some scientists. The abyss is imagined to be a biological desert—a flat, unchanging expanse of sediment stretching endlessly from one continent to another, broken only by the mysterious, linear mountain chains that bisect ocean basins and girdle the globe like seams on a softball.

That is wrong. The abyss is not a desert, it is not exclusively flat, it is not unchanging, it is not even universally cold and dark. And we know now that those mysterious mountain chains hold the key to a new understanding of how the earth's crust is formed and, possibly, where life on our planet originated.

The circumstances that delivered me to the gangway of my first oceanographic research expedition, and the avenue that led me to the professional career that had until then been the stuff of daydreams, were not straightforward. There are few universities where one can specialize in deep-sea research. I applied to the best of them for graduate school after earning my undergraduate degree in zoology from Rutgers University in 1977, but I was not accepted at my first choice—the Massachusetts Institute of Technology (MIT)/Woods Hole Oceanographic Institution Joint Program. Rather than settle for a second-choice school, I took off for North Carolina and a job as a technician. To learn new skills, I worked my way from one lab to another over much of a decade. I lived happily by the seaside in the company of others equally intent on drawing out some special secret of the animals and plants that live in tidal waters.

It was in Fort Pierce, Florida, that I found work at a Smithsonian field station that eventually opened a door to the deep sea. I was hired primarily on the strength of my ability to translate Russian papers for one of the scientists there, Robert Gore. Bob's specialty was the study of the larval stages of crabs, and I quickly worked my way toward assisting him with his science. The idea was to compare morphologies and look for natural groupings of species from the morphological details of the larvae. I spent hours watching larvae swim in small pools of sea-water beneath a dissecting microscope, fascinated by the trembling pulse of the heart and the undulating sweep of the gill bailer, both visible beneath a transparent exoskeleton dodged with ganglionic bursts of black pigment. I dissected preserved larvae into their component appendages and painstakingly traced each detail. I drew every one of the dozens of hairs to scale and accurately positioned them in a series of drawings made first in pencil and then carefully inked. I learned the technical names of the anatomy and could tell at a glance a maxilla from a maxilliped, an ischium from a merus, a basal endopodite from a proximal endite.

Eventually I left Florida, following a boyfriend to Cornell University, where he was to teach for nine months. I found myself at a critical juncture, unemployed, uncertain about where my life was going to take me, uncomfortable with the thought of being nothing more than a college professor's girlfriend. It was 1981, and hydrothermal vents on the seafloor—submarine geysers and their attendant faunas—had recently been discovered. Just a year earlier, Austin Williams, a senior scientist at the Smithsonian Institution, published a scientific description of a new crab species found at the vents. I wrote to Austin and asked if I could examine the eggs carried by some of the females in the collections stored at the Smithsonian. I thought that I might be able to remove the embryos and study their morphology, just as I had done with crab larvae when I worked with Bob Gore in Florida. Austin agreed and I began work on my first, very modest contribution to deep-sea research. When the material arrived—a few small vials with clumps of eggs looking like clusters of miniature grapes—it was as precious to me as any moon rock to a NASA scientist.

I learned that another major biological expedition to vents was to take place and asked if I could go along. There was room on one of the ships, and so I was able to join an expedition that carried all of the most respected researchers in the field of deep-sea biology—a scientific dream team.

I flew away from my boyfriend and a snowy Ithaca in mid-April to the balmy breezes of San Diego, and taxied my way to the dock to find the blue R/V *Melville,* a ship operated by Scripps Institution of Oceanography. She was the largest ship in a flotilla of three that would set sail for the East Pacific Rise. We were to occupy a single station for five weeks, with only a three-day port stop in Mazatlán, Mexico, midway through the cruise.

I was a timid soul as I marched up that gangway with my duffelbag on the eve of my first oceanographic expedition. I was unsure of myself but filled with anticipation. Now I am far more confident when I head off to sea, yet that sense of anticipation that makes my mind spin and my heart race has never gone away.

The real workhorse of our fleet—the submersible *Alvin*—was carried on the smallest vessel, *Lulu*, an ungainly catamaran that served as the mother ship for the submersible at that time. *Alvin* is an untethered vehicle, a wireless acoustic link being the only means of communication with the surface once she is submerged. She carries three people—one pilot and two observers—to depths sometimes greater than two miles beneath the surface of the sea.

Although I longed to dive in *Alvin*, I was content to be on one of the companion ships where the animals *Alvin* brought up from the vents were parceled out to waiting investigators and carried off to laboratories for experimentation. Giant tubeworms and clams were premium items, coveted by all of the science party and very dear. There was less interest in the crabs and squat lobsters, which I claimed as my niche and shared amicably with others. So I led a peaceable life, isolated from the intermittent scientific squabbles over who had first rights to which animals. It was a privilege just to see the animals and to attend the science meetings held each evening in the ship's library where that day's dive was reviewed and the next day's dive was planned.

I visited *Lulu* one of the days I was at sea, transferring from one ship to another on a small Zodiac, and swam in blue ocean water between her pontoons. As I slipped into the sea, I was instantly aware of the miles of water beneath me in a way that I had not appreciated when the firm steel deck of the ship was beneath my feet. In that first instant, I was frightened, although I quickly recovered and enjoyed my swim call. Mine was an innate response to an environment for which I am not adapted. It was a vivid reminder that descending the water column in a submarine is an unnatural act.

During that day on *Lulu*, I watched closely the men who tended *Alvin* and envied them the challenge of their job. I admired their experience, their confidence, their knowledge of how to work in the deep sea, their esprit de corps. When the submersible surfaced that day, I rode the small boat and handled the lines during *Alvin's* recovery. I dreamt then of just once being able to dive in *Alvin* as a scientist. I never dreamed that one day she would become my boat, in my command.

By the time the cruise was over, I knew I could never return to my old life. I was driven: I needed to know more about the seafloor and its ecology. I needed to understand the engineering and technology that enabled us to get to the seafloor and to make observations there. I needed to learn how to use that technology to answer questions of my own. Although I loved him, I left my boyfriend in Ithaca and took off on my own. I knew exactly where I wanted to go with my life, although I had little idea of how to get there.

My first step turned out to be graduate school at the University of California, Los Angeles, where I worked my way through undergraduate math courses, starting with calculus. Calculus was a subject I had dismally failed to learn in college, but under the tutelage of Professor Redheifer, a masterful teacher, calculus became an easy game. I progressed happily through differential equations and linear algebra to upper-level engineering courses on time-series analysis and computer programming. By the time I was through, I could integrate long strings of impressive-looking trigonometric functions and Greek symbols. I thought in terms of sines and cosines and natural logs punctuated by deltas, omegas, and psis. I could prove theorem after theorem in linear algebra. And I was finally material worthy of consideration by MIT. With a master's degree and a strong math background behind me, I reapplied to the MIT/Woods Hole graduate program and was accepted.

I secured my doctorate from MIT and the Woods Hole Oceanographic Institution by studying the ecology of some of the deep ocean's more exotic and interesting denizens. But the deep sea is a compelling place, and being just a passenger on an occasional dive to the seafloor did not satisfy. Becoming an *Alvin* pilot seemed to me a logical next step in my career as a deep-sea ecologist. Not only do pilots dive far more frequently and in more places than any scientist, they have the ultimate control over the dive. What better way to see the seafloor?

None of my colleagues agreed. Even Fred Grassle, my graduate advisor who supported me in all other ways, gently suggested I think long and hard about such a choice. There was a big risk involved—becoming a submersible pilot is too technical a track, not at all part of the usual career path in the academic world. Anonymous colleagues flatly stated in writing that they would refuse to dive with me if I qualified as a pilot, citing safety issues—could one really expect a scientist to operate the boat?

The day after I defended my Ph.D. dissertation, I started working for the *Alvin* Group. I knew it was the right thing to do, although it would take me once more off the beaten path to professional "success." I still do not have the academic position that colleagues with fewer credentials enjoy, but what I have gained is worth far more.

2

THE RIGHT STUFF

To become an alvin *pilot is unusual, and* little in my upbringing would seem to have set me off in such a direction. I come from a traditional American household. As I grew up through the late 1960s and 1970s, my parents remained locked into the role models and traditions of the 1940s and 1950s. Boys were to do boy things and girls were to do girl things. Each of my brothers had a workbench in my father's basement where they sat beside him with their tools, building things and tearing things apart after dinner or on rainy Sunday afternoons. The daughter of the family, I was to remain upstairs helping my mother clear the table and do the dishes or perhaps learning how to sew on the old Singer. I never had my own workbench, although I would often sit beside my father as he worked at his. I sought every opportunity to avoid women's chores, homework being the excuse that usually sufficed; I had to pay for my escape from the sink or dustcloth or vacuum by spending extra time in my room looking as if I were doing home-work. The one time I brought home a college boyfriend, Dad said after dinner, "Come on, son, let me show you my workshop while Cindy and her mother clean up the dishes." I flushed with teenage resentment of the female role I was supposed to assume.

I don't think I was a tomboy, but I remember my jealousy of my brothers, who each had an apple tree in the backyard. The trees, as if by legacy, came with full terri-torial climbing rights that my brothers assumed dictatori-ally. The two oak trees in the yard were assigned to my brothers as well. After I fussed and asked for a tree of my own, my brothers laughed at me when Mom and Dad suggested the cherry tree should be mine. The cherry was a pitiful tree, nearly in our neighbor's yard and with no good climbing branches. Moreover, it bloomed each spring with disgusting pink fluffy flowers that looked like wadded up toilet paper especially when it rained, and the foolish tree never bore fruit (in contrast to the deliciously sour albeit wormy green apples of my brothers' trees).

I probably never lacked my father's respect. But only when I was training to be a pilot did I finally begin to feel that I had earned it. By then he was retired from his job as an electronics technician for the government, and he was dying. We started to talk about things he surely had not expected his daughter to know about or be concerned with. We talked about the virtues and foibles of digital voltmeters, the puzzle of how a penetrator should be designed so as to pass circuits through the hull of a submersible while keeping water out, or how the complicated plumbing of *Alvin*'s variable ballast system worked. Dad recited with me the politically incorrect rhyme I'd been taught to keep track of resistor color codes: Bad (black) boys (brown) rape (red) our (orange) young (yellow) girls (green) but (blue) violet (violet) gives (gray) willingly (white). He had passed this on to my brothers when they were boys; now, in my late thirties, it was my turn. I told him of my satisfaction in having the right tool for the job, my simple pleasure in the split screwdriver that was my tool of choice when I had to work the terminal strips in one of the submersible's junction boxes. I told him of my pleasure in the challenge of troubleshooting and correcting a fault in the submersible when something went wrong a mile and a half below the surface.

Just before he died, my Dad, in a gruff and self-conciously offhand manner (a cover for emotion, I finally learned), told me to take his toolbox. His toolbox is a well-crafted wooden box that he had made of pine and carried to and from his Saturday job at Uncle Buck's radio repair shop in Red Bank, where he had worked for forty-odd years, since before I was born. Dad had other toolboxes that we were allowed to borrow from, but this one I remember as off-limits. It held his favorite tools. The box itself is delightful and complicated. Two doors swing open in front and hold sets of screwdrivers and nutdrivers, each in its own slot, and in a drawer that pulls out smoothly he kept small brown envelopes, each neatly labeled and filled with resistors. The top lid of the toolbox lifts up to expose a compartmented tray, beneath which is an area to hold larger pliers and crimping tools. It is far better than any workbench or some dumb old tree.

My Dad lived long enough to see me finish my training and qualify as a pilot. I think he was prouder of me for this accomplishment than anything else I have done. So am I. Training to be an *Alvin* pilot is not a trivial undertaking. The first hurdle is getting hired. I showed up at the right time with the right skill—the ability to put together documents and reports—just when the *Alvin* Group needed to develop a maintenance manual to continue its dive operations under navy certification.

I started work on the manual as *Alvin* arrived at Woods Hole in the winter of 1989 for one of her major overhauls, which take place every three to four years. Hoisted off the ship and laid up on the dock, *Alvin* is unbolted, unscrewed, stripped down to her smallest components. She is permitted no modesty. Her empty titanium sphere is set in a corner of the work space and inspected inch by inch for nascent failures or flaws. The skeletal titanium frame is shipped to New Jersey, where cracks are rewelded. The orange conning tower sits abandoned on the dock amid odds and ends of superstructure. Coils of cable and miles of wire are stowed in boxes on racks of shelves, instruments are lined up on work-benches for inspection and repair. Each poppet and spring is inspected, each valve rebuilt, each o-ring replaced. Paperwork spreads over every desk and countertop, covers every clipboard, fills dozens of manila folders. Signed and cross-checked inspection and repair forms accumulate as the teardown and rebuild progress. A handful of pilots and technicians and engineers do all the work. It was my job to follow along, turning lists into procedures, notes into tests, sketches into schematics.

When *Alvin* was back together again, I joined the seagoing crew as electrician's apprentice and pilot-in-training (PIT) on her first science leg, learning to charge her batteries and to troubleshoot and replace her cables. When the pilot who was also the electrician moved to a different job, I filled the electrician's post as best I could. I learned that not knowing how to do a job would not excuse me from that job. No one else was going to do it for me. I had to learn, quickly and proficiently. The job does not allow for insecurity, lack of confidence, or mistakes.

To become a pilot, I had to demonstrate a practical knowledge of *Alvin*'s systems, and an ability to deal calmly and safely with any emergency on the seafloor, independent of support from the surface if necessary. This kind of challenge suited me. I was well motivated and certainly used to studying and figuring out how things work.

That I barely knew how to use a wrench didn't bother me. I knew I was educable.

There is no formal PIT program at Woods Hole. One does not attend a series of classes and thus become a pilot. I learned on the job, at sea, working seven days a week, month after month. I was up with the rest of the group at 5:30 A.M. to get the sub ready for her 8:00 A.M. launch, working through a thirteen-page list of checks and procedures. While the sub was on the seafloor, I learned to track her from the surface using the acoustic navigation instruments. When the sub returned on deck, I was there for postdive checks and to help with work on the inevitable repairs that sometimes stretched into the next day. I qualified as a swimmer to handle the lines as the sub was lowered and picked up. I qualified as A-frame operator, following the instructions of the coordinator to pluck the sub from the sea without mishap in even the worst of weather. I qualified as coordinator and choreographed on the VHF radio the final check of the submersible systems and her launch and recovery.

When I wasn't working on *Alvin*, I studied her so hard I thought my brain was going to burst. I would fall asleep with my pencil in hand, sketching out one more time from memory the power distribution system of the sub, tracing the cabled route that lead from the big 120V batteries to the converters and the circuits that led to the control bus and equipment panels. When the sub was on deck, I would work inside her and, with my eyes shut, reach out to touch a specific one of the hundreds of toggle switches to learn their locations by heart.

Many nights my bunk was blanketed by blueprints as I worked my way through the schematics of the variable ballast system. My index finger traced the paths of fluids through its five-way valve, its autoclave and shuttle valves, all connected by a complicated plexus of plumbing, as I learned the logic of its control. Other times I pored over safety specifications, memorizing lists of requirements and emergency procedures. She did not escape me even in sleep. *Alvin* dreams filled my nights just as *Alvin* filled my days.

And every once in a while I dived as copilot, learning to handle the sub's systems on a science dive with a pilot looking over my shoulder. Those were the days that made it all worthwhile.

The length of the training period for a PIT is unspecified but usually takes a year at sea and at least half a dozen dives under the supervision of a pilot. Becoming a PIT does not automatically mean you will become a pilot. Some PITs take several years to qualify; others never make it. Most potential PITs quit within the first couple of weeks, unhappy with the stress and demands of the job or with the prospect of the required eight or nine months per year at sea on a small ship. Training to be an *Alvin* pilot was not easy for me. It was always intense and challenging. It was sometimes cruel.

When my learning was slow or I showed some particular lack of confidence in my ability, Dudley Foster, the senior *Alvin* pilot, would take me aside and give me just the bit of encouragement I needed. There was one worst time, though, when Dudley, a former jet-fighter pilot, took me aside one night on the fantail and told me he was not sure that I had the "Right Stuff." *Alvin* pilots, I knew, must thrive under pressure. I listened and vowed silently not to let this fresh cut in my self-confidence show. I demanded still more from myself. I was determined not to fail. At the outset, it did not occur to me that some of the pilots might never become supportive of my effort to join their rank. I knew some of them balked when they learned I had been hired as a PIT, but I was certain I would quickly earn their respect by hard work and my eagerness and ability to learn.

For the younger pilots, I had two strikes against me before I even began that were impossible to overcome.

First, and most damning in their minds, was the fact that I brought no skills to the group that would improve technical aspects of the submersible. Although the individuals who managed the operation appreciated the value of having a scientist among the rank of pilots, some of the guys I had to work closely with on a daily basis believed that as a scientist with no technical skills in electronics or mechanics, I had nothing to offer. In fact, I was explicitly told not to speak with scientists on the ship. Colleagues of mine reminisce about the preposterous situation of having to sneak into my stateroom in the evening, after my work with *Alvin* was over for the day, just to talk one sentence of science.

The second strike against me was my gender. There is no denying that the *Alvin* Group was, until I arrived, a man's world. Some of the *Alvin* pilots take great pride in the macho aspects of the job. I think my presence threatened the youngest men in that elite group, and I believe I was made to pay for this threat. That I succeeded in the face of their continued disapproval is due to the support from the more senior pilots—Ralph Hollis, who encouraged me to pursue the training and retired just as I began; Dudley Foster, who pushed me hard because he believed in me; and Pat Hickey, who has always been a friend.

I set about training with as low a profile as possible, but my very ability and drive to learn seemed to be a liability. Some individuals in the group were quick to blame every mistake on me, to ridicule my questions, to crow at my ignorance, even to give me false information. But because there were those who wanted me to fail, I felt compelled to succeed. In the end it was this compulsion that kept me from walking away. All of the good reasons that I had for wanting to pilot *Alvin* were no longer enough to make me put up with the abuse I felt. Proving that I could do it, and do it well, was finally my sole motivation when I began my exams. I might have quit even then, under the pressure of an examiner who wanted to see me fail, but I felt the burden of my gender and refused to give up.

I flew from my last qualifying boards in San Diego straight to the ship in Astoria, Oregon, where I joined the rotation of *Alvin* pilots and made my first "solo" dive, carrying two scientists in the sphere with me to a type of hydrothermal vent known as black smokers. Not long ago I reviewed the video record of that first dive and was impressed by how incredibly unskilled I was—safe, but unskilled—and how wonderfully patient the scientists were. Over the next year and a half I became a competent pilot, making forty-eight dives as pilot-in-command. I logged hundreds of hours on the seafloor during dives that took me to the middle of the Atlantic Ocean, to the Gulf of Mexico, and to the eastern Pacific, from off the coast of Vancouver south to the equator.

I made my last dive as a pilot in December 1991. Although I loved diving the boat, I knew my real career was in research. Ironically, to be a successful deep-sea scientist, one cannot spend nine months a year at sea diving on the seafloor. Now that I sail as scientist again, doing my own research, I am lucky if I get two or three dives in a year. More than once, the *Alvin* Group has asked me to return to my pilot's position. I am proud of the invitation and the hard-won respect that underlies it. So far I have declined, but always with regret that I cannot live two lives—scientist and pilot—at the same time.

3

ALVIN

EACHING THE SEAFLOOR IS A STUDY IN understatement. Although the technological feat seems akin to sending a man into orbit about the earth, *Alvin* dives up to 4,500 meters below the surface of the sea, day in and day out, following a routine that is stunningly anticlimactic. There is no countdown, no army of personnel to supervise the launch or recovery. Even the audience of curious scientists diminishes to naught after they have watched one or two launches; thereafter, they gather to watch only when the seas have built to a height they think might challenge the skills of the operators. The submersible's and ship's crews pride themselves on making the whole operation appear effortless. Although seemingly casual, the operations are highly choreographed yet responsive, with flexibility born of the legacy of maritime improvisation.

Alvin is clearly a boat that has worked hard for her living and even if I had never dived in her, I think I would know her to be special. In name she is thirty years old now, although little remains of her original body. Her exploits are legend. She was the first submersible to dive on the *Titanic,* which had lain undisturbed on the seabed for seventy-three years. Perhaps her most heroic and patriotic accomplishment (so far) occurred in the mid-1960s when her pilots and crew took her to waters off the coast of Spain, where she located the lost H-bomb from an air force B-52 that had collided with a refueling tanker in midair. In 1983 she made a series of dives on the remains of the *Thresher,* the nuclear submarine that had sunk twenty years earlier with all hands lost during her maiden sea trials. One time in 1968, more than twenty years before I first piloted *Alvin,* the steel cables used to raise and lower her cradle broke. The submersible, hatch open, pitched into the sea. Three men in the sub preparing to dive escaped, but *Alvin,* swamped by waves, filled with water and sank to the bottom in 1,535 meters of water. Dive number 308 lasted nearly a year, until she was salvaged by another sub, the *Aluminaut.* Over the years since then, her pieces and parts have been replaced and improved, from the smallest nuts and bolts to her skeletal frame and even the sphere itself. It would be difficult to identify a single piece of original equipment.

She is an ungainly water creature, streamlined only at her aft end, where she terminates in a snub fiberglass tail cone and a stainless steel towing bridle. Forward, her Neanderthal brow juts out, spiked with lights and cameras. A battered and burned work basket projects low down in front like a bumper. Manipulators, fed by red veins of hydraulic fluid, fold elbows-up beside her cheeks like legs of a praying mantis. Her sides bulge out like the

DSV ALVIN

flanks of a fattened pig. Perhaps we are fond of her just because she is so ugly and utilitarian.

A portrait of *Alvin* sits above the couch in my living room. The artist has painted her submerged, sinking through the water. Last bubbles of air ballast escape from the vent valve on her sail to stream to the surface. Floating in the water above her, I have watched *Alvin* dive. My eyes lock on her; it is impossible not to track her as she goes down. Although I know the routine intimately by now, it still seems an unnatural thing to watch her sink farther and farther in the clear ocean water until she becomes just a tiny speck and then disappears altogether from my sight.

From inside the sphere, the surface of the sea, clear and faceted by shafts of sunlight, flashes in front of the viewports. As the descent begins, clear water quickly becomes aqua, then a deeper blue-green-black that has no name, then darker still, until there is no color left at all, only blackness. In this colorless transparency of water 1,000 meters and deeper, splashes of bioluminescent light silently pass by like shooting stars; they are the only index of motion as the submersible continues its plunge.

Redness lights the interior of the submersible; the glow from dials and video screens reflects softly off the burnished titanium hull. Racks of black instrument panels lined with banks of silver toggle switches surround the pilot. Sensors and gauges and meters silently monitor the progress of the dive. Black cushions cover the metal deck, low in the sphere and sloping down to port and starboard, where, farther aft than the pilot, two scientists sit opposite, legs tangled across the short cord. They share their space with lunch boxes, thermoses, cameras, and laptop computers, CO_2 absorbent, blankets and sleeping bags, fire extinguishers, flashlights, and personal gear.

In the tropics, the submersible sphere is a sweatbox on the surface, but as the dive progresses, a penetrating damp chill invades. The cold water of the deep sea is ancient, and the chill that passes through the hull seems old, heavy, relentless. Moisture collects on the inside surface of the hull, beading and streaking to form rivulets that course down the sides of the sphere to soak into cushions and clothes.

Like a metronome, every five seconds the submersible generates an acoustic tracking pulse at 8.1 kilohertz. The signal sounds like the puck of a tennis ball hit hard against a racket. Every fifteen seconds, the navigation system sends out a query at 7.5 kilohertz. Returns from moored acoustic responders answer back, and a volley of pings and echoes at signature frequencies is heard in rapid, syncopated rhythm.

Pilots listen for the continuous hiss of oxygen flowing into the sphere, the grinding hum of the CO_2 scrubber. There is the thunk of a valve seating, the static of the hydraulic plant, the whine of the thrusters, the throaty gurgle of the variable ballast pump as it works to push water out of holding tanks against the pressure of thousands of meters of water.

At intervals, a Morse code signal is keyed by the surface controller on the ship; a shorthand message is passed, requesting the submersible's depth. The pilot keys the underwater mike and responds. The exchange is sweet and regular assurance that all is well, top and bottom.

Alvin carries her three-person crew in the titanium sphere at the forward end of the vessel. Four inches thick at the viewports and hatch, the sphere can withstand the pressure of more than 4,500 meters of overlying water, a pressure so far beyond my comprehension that I do not even try to relate to it in any way other than in engineering terms.

Cozy is a generous description of the inside of the sphere. It looks a lot like the cockpit of a jet, at least in the degree to which you are engulfed by instruments. There is an obvious lack of comfortable seats. The pilot gets a small cushioned box to sit on, the pilot's privilege. Observers have to lounge—a generous verb for contorted limbs that are shifted about continuously in the perpetual search for even marginal comfort—on the cushioned deck, all 6 square feet of it. Their combined allotted space is not much bigger than a standard bathtub. I receive many compliments on my small size from long-legged and big-bodied scientists who dive with me and are happy to have the extra room.

Each observer has a small viewport that permits a framed and precious view of the submarine world. These viewports are low in the sphere, designed for close and thoughtful observations of the seafloor. The view is worth every tortured moment of discomfort it takes to hunker down, scrunch up, and peer out. The best view is often out the pilot's viewport, higher up, when the pilot works a vertical surface like the face of a cliff. Then the scientists watch the video monitors that display the forward-looking camera images, since out their viewports there is nothing but water. I was spoiled by my prime seat as a pilot and remember fondly that view when I dive now as a scientist.

Drab-olive military packages of emergency rations and water are tucked into a back corner of the sphere. The lunch packed daily by the steward is pretty basic. Whereas the French submersible *Nautile* is reported to pack wine and four-course hot lunches for her passengers, meat or peanut-butter-and-jelly sandwiches are standard *Alvin* fare, stowed before the dive together with a candy bar, a piece of fruit, and coffee. This picnic lunch is usually eaten in random bites on the seafloor.

There are no special requirements to be a passenger in *Alvin,* although you must be able to fit through the hatch and be of sound mind and reasonably sound body. Before your first dive, an *Alvin* crew member takes you barefoot into the sphere while the sub is on deck for a safety briefing. Once you are settled inside, *Alvin* basics are reviewed and your response to the tight quarters is observed. I have never dived with anyone who seemed overtly uncomfortable in the sardine-can confines of the sphere, and only once did a person I was briefing opt not to dive because he felt claustrophobic.

The part of the briefing I like best is the *Operator's Manual*. It is a white, three-ring binder that provides a terse summary of how, as the name indicates, to operate *Alvin*—as if this were a Chevrolet from the car dealer down the street. The *Operator's Manual* is stowed on a shelf in the sphere, dubious reassurance for a passenger with an incapacitated pilot. In fact, it is not so unreasonable a guide, since there are seventy-two hours of emergency life support capability, which ought to be long enough for a passenger to study the manual and confer with the ship by underwater telephone so that the sub can be brought safely back to the surface.

Unlike SCUBA divers, who dive under ambient water pressures and have to decompress on their way to the surface after a deep dive, passengers in *Alvin* stay comfortably at one atmosphere, or sea-level pressure. Air quality is maintained by bleeding hospital-grade oxygen into the sphere from a tank and removing respired carbon dioxide using a chemical absorbent. Three bottles of oxygen and four cans of absorbent provide those seventy-two hours of emergency life support.

Every passenger preparing to dive in *Alvin* is shown how to bring her back to the surface. She is designed expressly so that leaving the seafloor is one of the easiest things she can do. *Alvin* sinks because she carries disposable stacks of weights—steel ballast—hung on either side, just aft of the observers' viewports. To leave the bottom under normal conditions, you simply have to drop those weights and *Alvin* will float up on her own.

The ultimate safety factor in *Alvin,* should all other attempts to raise her off the bottom fail at the seventieth hour of an emergency, is the release of her titanium sphere. The sphere with her passengers is packaged with sufficient buoyancy material to send her racing to the surface once she is cut loose from the afterbody. We are proud that this extreme safety measure has never had to be used.

Once *Alvin* is submerged, communication between *Alvin* and the surface is through the underwater telephone, or UQC. This system is what makes diving in a submersible sound like you are diving in a submersible. Usually the communication is as clear as a phone line, except for the echoes that sometimes roll on top of one another. On a calm and rainy day, if you listen closely as you ascend, you can hear the rain before you reach the surface—a light pattering sound. You can hear the engines of the ship and the outboard motor of the small boat that is standing by to assist in the recovery. Occasionally you hear whales and dolphins whistle and, much more rarely, pings and chatter from another submersible somewhere in the vicinity.

The ordinary cues for traveling, the visual sense of distance and motion, are absent during the descent to the seafloor. But sinking like a stone at a terminal velocity of 30 meters a minute, the pilot has to keep track of the bottom. At the launch, the surface controller gives you a launch altitude at the same time that permission to dive is granted, so you know the general depth of the water beneath you when you submerge. Pressure transducers with digital readouts let you keep track of your depth and altitude off the bottom as you descend. You still must monitor your altitude above the bottom so that you can control your bottom approach by adjusting your ballast. *Alvin* is a tough boat, but she does not take too well to crash landings on the seafloor. I don't know this from personal experience. But pilots who have lost track of the seafloor can't hide the fact, because hitting the seafloor at 30 meters a minute unnerves the scientists in the sub, and they are quick to announce the event when they return to the surface, much to the chagrin of the pilot.

When I first started working with *Alvin*, I expected her gear to be high tech and digital, and I felt a curious disillusionment that anything as anachronistic as a windup stopwatch, used as a backup for timing acoustic signals, should be a standard piece of gear in the sub. In fact, all of the critical sensors on the sub have analog or nonelectronic backups to the digital readouts, and I have more than once been glad to have them. Digital readouts are fed into a datalogger on the sub's computer and relayed to a video monitor. I have made at least one dive during which the computer crashed and I had to work the entire dive with my analog outputs.

Alvin is a safe boat, with an unsurpassed safety record. Even so, no pilot can work the seafloor without wondering what it would be like to find oneself trapped 4,000 meters down with a short seventy-two hours to sort things out. Each one of us has been in places we would rather not have been. But only once and only briefly did I find myself actually afraid I could die on the seafloor. I spooked myself with a fleet but terrifying thought of being buried alive by a "landslide" while I was working down in a narrow fissure. My heart raced, my fingers trembled, my imaginings of a slow and hopeless dying ran wildly unchecked for one instant, and I wanted only to fly up and out of that fearsome crack in the seafloor that moments ago I had so admired and deftly maneuvered through. None of *Alvin*'s safety features would do me any good if she sat beneath a ton of rock. I hit the "up" switch on the joystick and precipitously put an end to the scientists' close observations in that fissure. I did not hear any challenge to my action.

In quiet talks among pilots I have since found we all share two great fears: getting trapped in a cave and being buried. For much of the other stuff that can go wrong, we know we have a way out. Operator error—either a pilot's or one of the crew that helps to maintain the sub—is avoidable by careful work with attention to detail and the rigorous schedule of checks upon checks that precedes every dive.

This is not to say that there have not been tense moments in the sub. But when a dial, meter, alarm, or unusual sound indicates a problem, you learn to mutter under your breath, work through switches to isolate the problem—electrical are the most common—pause a moment to gather your facts, and if it is really serious, call up on the UQC with a calm assessment of the problem. I have had to abort a dive only a few times because of a critical failure I could not circumvent.

One of my aborted dives earned me the nickname "Scoop," born of the several hundred pounds of mud that slipped into the underbelly of the sub, between two fiberglass plates that ought to have been buckled together but were not, as I drove up a shallow rise off the Oregon coast.

I was working an area of mud dunes, a curious region where mud hills rise gently and fall off abruptly. The vista is of a relief that is softly imposing, with smoothed contours like the dunes of the Outer Banks, colored in earthy dun and sienna. I had been staying close to the bottom to hunt for clams, but decided it was time to rise up and fly higher in a different direction. I toggled up on the thrusters, but I couldn't get off the bottom no matter how much power I put to the lift props. Although I had no reason to suspect I had snagged on something, I also had no reason to suspect I had ballasted myself with mud. If I had snagged, I was in no hurry to drop my weights, since that could put me in an even worse situation.

Over the next hour or so I worked carefully to figure out what was going on, staying in touch with the surface controller. I was perplexed but not worried. We did finally leave the bottom and ascend very slowly. As we passed through the thermocline at about 500 meters I knew exactly what had happened, since the temperature probe that sits in the belly of the sub stayed at a chill 5°C— insulated from the warm water by a load of abyssal mud. Once we were back on deck, it took more than an hour to wash the mud from *Alvin*'s belly.

An inherent potential for danger is involved in sending a manned vehicle to the seafloor, an argument used to win support for the development of unmanned tethered vehicles. There is no reason why a robot equipped with multiple cameras and manipulators could not accomplish the tasks *Alvin* can do. But so far, despite much money put into development of unmanned systems, *Alvin* has not become obsolete. Her strengths are her experienced pilots and her well-tested technology, which make her a reliable workhorse.

I would hate to be totally excluded from visiting the seafloor myself. There is an undefinable advantage to seeing the ocean bed with one's own eyes. I think a creative force and a passion develop that are unattainable from watching a mere image, no matter how good that image might be. As a colleague of mine once pointed out, no one who has a choice between watching a video of Paris or going there in person is going to opt for the armchair approach. The same holds true for the seafloor. I am more than willing to take a certain risk to get there.

I like the details of how *Alvin* works. Most aspects of her operation are stunningly simple and elegant. There just happen to be a lot of simple aspects that combine to make a complex whole. The fundamental principle is that you sink because you make yourself heavy, you surface because you make yourself light; you get neutral by adjusting your ballast water until you float like a jellyfish. No pilot dives the boat the same way as any other. What one pilot holds as sacrosanct routine, another pilot disdains. I worked out my own routine of reminders and checks and ways of getting things done. As a scientist, you learn to work with each pilot's idiosyncrasies and to take advantage of strengths. *Alvin* is the only deep-diving submersible that routinely works with a single pilot while carrying two observers. Diving "solo" is a matter of pride among *Alvin* pilots.

At sea, the rotation of pilots follows a strict order so that no one pilot dives more than another. During lean cruises, when the number of pilots on the ship is down to two, this means that every other day you are on the seafloor and during your off day you are surface controller. In your spare time at night, when the boat is on deck, you take care of whatever repairs need to be done so she can dive again the next morning. Sea duty for a member of the *Alvin* Group is at least eight months a year, broken up into two and a half or three months at sea, with about a month off in between. Vacation time is all too short. One is barely settled into shore life—bills paid, friends and family visited, business taken care of—when it is time to disappear to sea for another couple of months. It is no surprise that at this pace, many pilots burn out after just a few years.

Once on station, the tempo at sea is an odd one. There are no weekends. The ship works around-the-clock regardless of the day, the submersible dives day in and day out, weekends and holidays. The progress of time is paced by weekly fire and boat drills and the countdown of dive days remaining. Saturdays might be marked with a barbeque and a beer, if the weather is gentle enough and the captain is feeling merry. Holidays are the most difficult time. The last Christmas I spent at sea, a string of miniature lights blinked back silver tears on a Charlie Brown fir tree that decorated the ship's lounge. One bough, weighted with a single red bauble intended for a branch of far more substance, gave the tree a decided list to starboard. The little tree, a gift of the ship's chandler, was a poignant measure of the loneliness of a holiday at sea.

No one takes the job for the luxury accommodations. Most of the guys in the *Alvin* Group have to share a room. Being the only woman in the group, I at least had a stateroom to myself. My Snake-Pit room was below the waterline; the port hull was all that separated me from the sea.

Little about the stateroom was nautical. It was bare of teak and lacked any hint of a brass fitting. Functional is the kindest descriptor I can apply. Two aluminum bunks hung on chains secured to sheet metal overhead and anchored to the concrete deck. Sheet-metal bulkheads encased the cubicle, and metal footlockers served as the only storage space. Duct tape kept the doors of the lockers from slamming with each roll of the ship. A pair of life jackets and survival suits were stashed beneath the bottom bunk. My computer was tied down to a metal desk. There was a stainless-steel sink, and a mirrored medicine cabinet. A comfortable chair, a rug on the deck, and a favorite picture made it my home.

Without a porthole, I would try to gauge the weather and the status of the ship each morning by the sound of water against the hull and the motion of the ship. I learned the sounds that told me whether we were steaming, or maneuvering, or on station—the slap of the water against the hull, the strumming of the engines, the frictioned complaint of the winch. Port to the dock, I could hear the squeal of rubber bumpers chafing against wood pilings and the shriek of lines as they tugged against bollards, tide and surge working the ship up and down. Only in the most extreme of seas could I judge the weather well enough to know that the day's dive would be canceled. If my books flew off the shelves and onto the deck as I was getting up, and if I had to hold onto the sink while I brushed my teeth, these were certain indications that a message on the board in the *Alvin* shop would tell me the dive was canceled and send me back to my bunk. Weathered-out days are rare for the sub. *Alvin* can be launched and recovered in difficult seas, and her schedule keeps her in appropriate parts of the ocean through the year so she encounters as few rough seas as possible. We did have to move offshore once when a hurricane pre-empted our dive site in the North Atlantic, and nasty weather does pile up off the Oregon coast, bad enough to keep the sub in the hangar and the chief scientist on edge. But generally we are fair-weather sailors.

Ironically, I do not make a very good sailor. The only knot I know is a bowline. I can find the Big Dipper, but the North Star can be elusive. I cannot throw a line. And I know that, fair weather or not, as the ship pulls away from the dock I will soon be seasick, a malady I suffer much to the amusement of my *Alvin* compatriots. It takes a day at least for me to become accustomed to the motion of the ship. In rough seas, the *AII* has a snap to her roll that is unsettling. I've learned to work through the nausea when I have to, but when I can slip away, I disappear to the Snake-Pit to curl up in my bunk and suffer silently and patiently. Yet I would have endured far worse conditions to pilot *Alvin*.

4

THE ABYSSAL
WILDERNESS

THE SEAFLOOR CAN BE SURPRISINGLY RICH IN visual textures. Where the ocean crust is young, lava flows dominate the landscape. The lavas pool and ripple and swirl in frozen motion. Pillows of lava with elephant-hide skins drape the slopes of submarine mountains like icing run down the side of a cake. Stilled lavas are torn, ripped apart, prelude and aftermath of the violent birth of new seafloor.

I have driven into fissures cut deep into lavas, deep beyond seeing, and followed them until their steep-sided walls begin to close in on either side. Exposed in the cuts are the histories of eruptions, flows built upon flows. Each flow shows as a discrete unit, the events and the

passage of time marked by layered differences in form and color.

Elsewhere, ponds emptied of lava with walls marked by bathtub rings are features common to the central valleys of submarine mountain ranges, but not found in any terrestrial environment. The bathtub rings are presumed to be lithic signatures of repeated filling and drainback of lava. Where particularly fluid lavas fill a depression, the surface exposed to seawater cools and solidifies, forming a solid rind an inch or two thick. The underlying molten lava drains back into the crust. The roof, now unsupported, collapses, leaving behind a level mark that rings the depression. Later, lava again fills the depression, again a crust is formed, the lava drains, the roof collapses, and another mark rings the emptied pond. Over and over the cycle repeats itself until a hundred or more rings accumulate. These collapse features can be extensive, big enough for the submersible to explore. I work in them often and all but take them for granted, although few people have ever seen them.

It is a silent world. I have set the submersible on the seabed and systematically shut down all of the systems. We call it "going dead boat." Silence and darkness are immediate and ultimate, all-encompassing and pervasive. I feel the silence more than hear it; it feels cold, oppressive, alien. My voice in the silence sounds thin and nervous, insignificant. Never am I more conscious of the tons of water that surround me. Only when I return power to the boat does my pulse return to normal.

It would be only a short hike—barely a few tens of miles—to cover the distance that submersibles have explored along the 46,000-mile extent of the mid-ocean ridge system. These ridges can look like highways, paved black with basaltic lavas. In some regions, the seafloor drops away 20 to 50 meters along the centerline of the ridge to form a rugged axial valley that might be 100 meters or more in width. The most recent lavas are found in this axial valley. They are glassy, reflective, with a rind of brittle glass that hasn't been in place long enough to weather or to accumulate a covering of dusty sediment rained down from the surface. The terrain in the axial valley can be tortured and rugged, filled with pits and caverns and tall pillars, with rubble, talus, scree. It is a landscape of stark beauty.

The lavas, their distribution and form and composition and age, provide clues to the dynamics and mechanisms of the geological processes that create the seafloor. They are the cache that geologists seek, and must be carefully plucked from the ocean bed. Sometimes the lavas are hard and dense, broken but in place and easy to pick up with the manipulator. Other times the lavas are intractable and must be left attached to the planet. Most often, the lavas are glassy and brittle, easily shattered as the manipulator's jaws close upon them.

Corals on top of seamount
(abundance
decreases downslope)

Branches of corals are
covered with flower-like
polyps.

Polyps embedded in
calcareous matrix of
coral branch.

As on land, local topography plays an important role in affecting the distribution of organisms. Exploring the summits of submarine mountains, I have encountered inverse "timberlines"—only the peaks were populated by stands of shrublike corals. The corals feed passively on bits of organic material suspended in the water. Only at the peaks is the current fast enough to supply food that sustains the corals.

In other places, natural controls on the distribution of animals are less obvious. Off the coast of California, on the soft mud bottom of San Nicholas Basin, red shrimp, jumbo sized, are the most abundant and ubiquitous of the large animals. When strong winds forced us to move our dive operation north and east a few miles to the lee of Santa Cruz Island, we dove in Santa Cruz Basin. There, elegant sea pens populate the mud bottom, their fleshy bodies creamy yellow, their plumes white and frilly. They are everywhere, a meadow of sea pens, all oriented into the current, ready to snatch small bits of life and debris as it passes by in the sluggish bottom tides. Their spacing is regular, a telltale mark that the flow field around each sea pen defines the optimum spacing of neighbors. We don't yet know why two adjacent basins support such contrasting faunas.

I have encountered herds of sea cucumbers. Picture your garden bed of cucumbers, a very large patch of them, a bumper crop. Take away the vines and expose the cukes, all lined up in one direction. Of course, deep-sea cucumbers are not vegetable but animal; they are related to sea stars, although not as aesthetically appealing. As I drive over the herd with the submersible, they roll out of the way like tumbleweed before the wind. The sea cucumbers are neutrally buoyant, made up almost entirely of water. They are transparent, except for the pentium of long bands of white muscle that run the length of their bodies at intervals. I do not know why they aggregate as they do, although food or sex are two likely explanations.

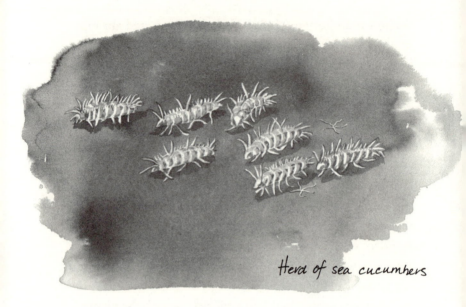

Herd of sea cucumbers

How fast they move across the seafloor, how long they stay together as a group, what effect they have on the sediment and the animals living in it are all unknowns.

Once *Alvin* is back on deck, I always go to her sample basket to touch the rocks I have collected, to feel how the rock looks. Rocks brought to the surface don't seem so extraordinary; in fact, to an untrained eye they look pretty much alike. But each is labeled and cataloged, chipped, cut, powdered, assayed, and probed. Once the rocks are grouped into mineralogical classes, the sequence of events that may have led to their formation can be deciphered.

Animals retrieved from the seafloor attract the most attention on the ship. I have seen knots of sea spiders come back alive and squirming in the bottom of the collection box; I have reached for slimy sea anemones that slipped out of my hand like soap; I have felt sorry for the captured ugly fish, all wrinkled and squinty-eyed and dead. Each is carried off to the laboratory for observation, description, experimentation. A single animal may be dissected down to organs and tissues, and then parceled out like gold dust to waiting biologists.

Where the seafloor is spreading apart, where it is rent by earthquakes, and where the heat of molten magma lies close to the surface of the ocean crust, seawater percolates through cracks in the basalt, down to where it is heated and reacts with hot rock. The chemically modified water, itself now hot and buoyant, channelizes through conduits in the seafloor to exit as hot springs within the axial valleys of submarine spreading centers. These hot springs easily rival their terrestrial analogs in power and spectacle. Pressure keeps the hot water from steaming or boiling; it becomes superheated, reaching temperatures of 350°C and more. Venting water, emerging clear from the seafloor, quickly turns into turbulent plumes of "black smoke" as dissolved minerals form particles on mixing with seawater.

Where the plumbing is leaky or the subsurface heat source is nearly cooled or quenched, vent water flows out of the seafloor as diffuse plumes and at warm temperatures (~2 to 30–40°C) more favorable to life than the high temperature extremes of black smokers. Vent water is enriched in reduced chemical compounds, especially hydrogen sulfide. A sample bottle of vent water opened in the laboratory can clear a room in seconds as the ripe odor of rotten egg escapes. A variety of bacteria thrive on the sulfide, using its chemical energy through chemosynthesis in much the same way that plants use energy from light to produce organic carbon through photosynthesis.

The stunning implication is that submarine hydrothermal systems, fueled by the heat of volcanic processes, can support life in the absence of sunlight. Vent water may be the ultimate soup in the sorcerer's kettle. The water has a primeval chemistry that has prevailed along submarine mountain ranges since the breakup of Gondwanaland. There is a growing interest in and legitimization of theories that the chemical and thermal conditions found in some vent waters may permit the synthesis of organic compounds. Deep-sea vents may have been the site where life originated on this planet.

The potential significance of submarine hydrothermal processes on the physical nature of our planet is impressive. Chemical reactions that take place in the subsurface plumbing of the seafloor may have, over eons, been a major influence on the elemental composition of present-day seawater. The thermal input from hydrothermal springs along submarine spreading centers may drive major patterns of deepwater circulation. And, at some sites, where hydrothermal activity has persisted for long periods, great mounds of metallic ores—iron, copper, and zinc sulfides—have accumulated. Many of the terrestrial commercial ore deposits are now understood to have originated on the seafloor as a consequence of the chemistry between seawater and hot basalt.

Entire communities of invertebrates have adapted to life at vents. Newly described species of clams and mussels depend on symbiotic, chemosynthetic bacteria for their nutrition. Productivity of vent communities is comparable to that of shallow-water coral reefs and salt marshes; the tempo of life is equally fast paced, with vent organisms growing at rates equivalent to their shallow-water analogues. In addition to the bacteria-invertebrate symbiotic associations, hundreds of new species of polychaete worms, crustaceans, and other invertebrates inhabit vents, where they occupy conventional niches of grazers and predators and scavengers within the community.

Name-giving is one of the perquisites of leading exploratory dives to vent sites. The first vent fields ever discovered in the deep sea—Rose Garden, Garden of Eden, and East of Eden on the Galapagos Spreading Center, just north of the equator—were named for their spectacular gardenlike displays of giant tubeworms. Clam Acres, explored by biologists in 1982, is a field of lobular lavas, each lobe necklaced by the white shells of giant clams.

Lucky Strike, Broken Spur, and Snake-Pit give a western motif to the names of major vent sites on the Mid-Atlantic Ridge. Lucky Strike was just that: a dredge brought up a chunk of fresh sulfide colonized by mussels and with that stroke of luck another hydrothermal system on the Mid-Atlantic Ridge was discovered. Broken Spur earned its name from its morphological distinction as a short spur of ridge that projects off a larger feature and is split down its axis by a valley. The Statue of Liberty and the Eiffel Tower at Lucky Strike celebrate the French-American collaborations that supported the first exploratory dive series there. TAG is the more prosaic name for Trans-Atlantic Geotraverse, in keeping perhaps with the nonfrivolous, bureaucratic milieu of the agency that sponsored much of the early exploration to that site. Relict mounds at TAG, called *Mir* and *Alvin,* are named for the submersibles that undertook the first major reconnaissance at each site. The *ARGO-2* fault runs beside the active TAG mound, an honorific for the team that winched the *ARGO-2* tethered camera system up and down the steep fault scarp without a single disastrous contact between sled and hard rock.

Often the features of a site suggest a name. The Spire at Broken Spur is a tall and spindly sulfide structure. Geologists named Snake-Pit for its abundance of white, eel-like fish that slither about over the surface of the sulfides. At the Beehive site of Snake-Pit and the Wasp's Nest of Broken Spur, the sulfide structures have bulbous, shingled features from which hot water flows. The Kremlin area at TAG is covered with sulfide features that are reminiscent of the onion domes of Saint Peter's Basilica. The Saracen's Head, though, was named by the

Black smoke vents through finger-like chimneys at the top of the structure

Saracen's Head

scientist who discovered it as much for a British pub he frequented as for the peculiar shape of the sulfides.

At Hole-to-Hell, lava erupted on the seafloor in 1991, creating a miasmic cloud of impenetrable black smoke that has since given way to black smoker chimneys and low-temperature venting. Virgin Mound on Axial Seamount is an inhospitable white spring that does not support life. Bob Hessler named Alice Springs and Burke for his two dogs. (You don't want to know how Dead Dog, a site at Middle Valley on the Juan de Fuca Ridge, got its name.)

The Dudley sulfide edifice at the Endeavour Hydrothermal Field is where a faithfully reproduced, life-sized plywood manikin of *Alvin* pilot Dudley Foster stands to serve as a reference for scale. Originally dressed in corduroys and an *Alvin* T-shirt but last reported stripped naked by bottom currents, the edifice towers over Dudley's 5 feet 8 inches.

Biologists get to name all of the new species. *Alvin* is a popular name. Among the polychaete worms, there are the genera *Alvinella* and *Paralvinella;* there is a snail *Alviniconcha* and a shrimp *Alvinocaris.* One tubeworm is known as *Oasisia alvinae,* the *Oasisia* referring to the name of the expedition—Oasis—that recovered the first specimen. *Alvini, alvinus,* and *alvinae* are all specific homages to *Alvin.*

Some new species are named for pilots (*fosteri, tibbetsi, hollisi* [limpets]), or scientists (*hessleri, grasslei, sandersi,*) or their dogs (*burkensis* [polynoid polychaetes]). Bashfully I submit that a dancing shrimp—*Chorocaris*—carries my name, *vandoverae,* an appellation assigned to the beast by my friends and colleagues Jody Martin and Robert Hessler. Other species' names, like *jerichonana,*

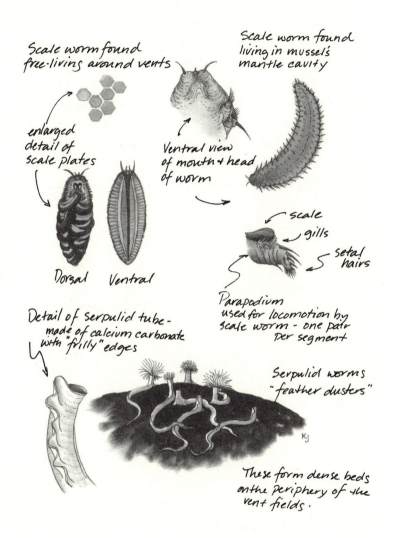

Scale worm found free-living around vents

enlarged detail of scale plates

Dorsal   Ventral

Scale worm found living in mussel's mantle cavity

Ventral view of mouth + head of worm

scale
gills
setal hairs

Parapodium used for locomotion by scale worm - one pair per segment

Detail of serpulid tube - made of calcium carbonate with "frilly" edges

Serpulid worms "feather dusters"

These form dense beds on the periphery of the vent fields.

vulcani, caldariensis, and *pompejana,* evoke the inhospitable habitats of hydrothermal systems.

Find a new vent site and you are guaranteed to find new species. Even returning to old sites, it is easy to discover new types of animals in myriad unsampled microhabitats. While tubeworms, bivalves, shrimps, and other large invertebrates attract the most attention, the real diversity of life at vents is found among the smaller invertebrates. A group of copepods, the siphonostomes, has been especially successful at invading and speciating

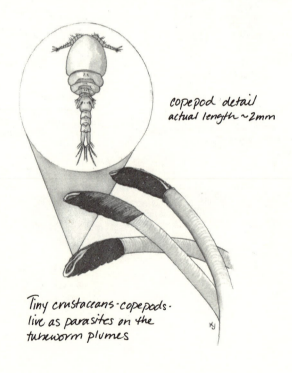

copepod detail
actual length ~2mm

Tiny crustaceans-copepods.
live as parasites on the
tubeworm plumes

within the vent habitat. Dozens of new species have been described by Arthur Humes, a retired zoologist who once taught me invertebrate biology and who continues to work full days at his microscope and manuscripts in a small office in Woods Hole. His copepods are tiny—flea-sized—and are discriminated by differing counts of hairs and lobes on appendages. They have sucking mouth parts—hence the name siphonostomes—and are thought to make their living as ectoparasites on the larger inverte-brates. I have seen them infesting plumes of tubeworms, crowding gills of shrimp, assaulting backs of crabs, a dif-ferent species of copepod for each host. There is even a red copepod that specializes on the fluffy giant filamen-tous bacteria found at Guaymas Basin vents in the Gulf of California, the copepods sprinkled like paprika over the white surface of the mats.

Polychaete worms are another group that have found diverse niches within the vent community. Some of the worms look familiar, allied to shallow-water species. There is, for example, a type of rag worm, or nereid poly-chaete, known from Pacific vents that belongs to the same group of worms that I have used to bait bluefish in the waters off Cape Cod. But then there are other poly-chaetes, an entire family of them known as alvinellids, that are so far found only at vents in the Pacific. Certain alvinellid species live very conspicuous lives in some of the warmest habitable waters of black smoker chimneys. These worms are red and fat, as big around as my thumb

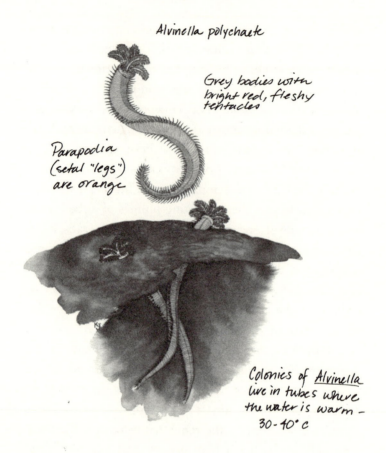

Alvinella polychaete

Grey bodies with bright red, fleshy tentacles

Parapodia (setal "legs") are orange

Colonies of <u>Alvinella</u> live in tubes where the water is warm – 30 - 40° C

and half a foot long. They live in tubes, their bodies are covered with a white fur of bacteria, their heads are dressed with a starburst plume of stout tentacles used to sense their world and to gather food. In the scientific scramble for superlatives, *Alvinella pompejana*, the pompeii worm, has been nominated as the animal, among the

64

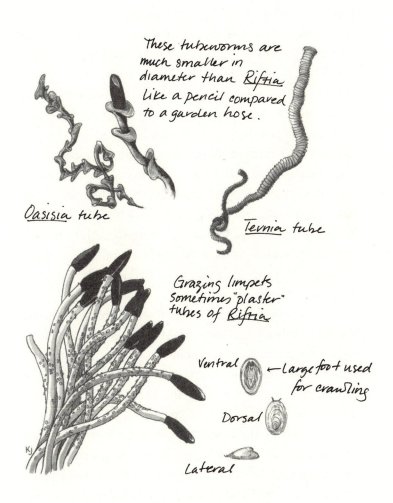

These tubeworms are much smaller in diameter than _Riftia_ Like a pencil compared to a garden hose.

Oasisia tube

Tevnia tube

Grazing limpets sometimes "plaster" tubes of _Riftia_

Ventral ←Large foot used for crawling

Dorsal

Lateral

entire animal kingdom, that lives in the hottest environment, routinely bathed by 40°C waters.

Limpets are likely to vie for status as one of the most abundant types of animals found at some vents. These small, untwisted gastropods plaster the tubes of tubeworms in some places; elsewhere their populations are so dense, with individuals piled one on top of another, that they can be scooped up in handfuls like pebbles. Squat lobsters seem to find the immature limpets a delicacy. I found the remains of more than forty shells in the distended stomach of one squat lobster and had an instant vision of this animal eating the limpets like popcorn. The limpets are a diverse group, too. Representatives of new species from vents around the world are piling up on the shelves of Swedish biologist Anders Warèn, who has made this group his specialty.

In my own rambles around vents I have stumbled upon and helped describe a few of the new species. I don't have a particular specialty group. If I find a new limpet, I send it to Warèn, new copepods I pass off to Dr. Humes. But for the peculiar vent bugs that aren't under the care of a particular specialist, I try to find taxonomic homes. The most unlikely of these orphans whose description I helped nurture is a new type of fungus that lives as a commensal in the stomachs of those same squat lobsters that feed on limpets. I am safe in claiming that the description of this fungus establishes a new depth record for the entire group to which it belongs. Another curious organism for which I helped find a taxonomic home is a single-celled foramini-feran, or protozoan, a small bit of protoplasm that gathers about it an agglutinated spiral of black glass shards. I found the protozoan attracted in large numbers to slate panels we deliberately left at vents for one year and then recovered. Among my colleagues, the protozoan seems to be an all-but-forgotten species, like the fungus, falling outside the scope of our ecological inventories.

From vents on the East Pacific Rise, I collected a more uncommon kind of new species that lives an unusual lifestyle: an amphipod, a type of crustacean better known as a beach hopper or sand flea. The peculiarity that fasci-nated me was its habit of living in swarms over warm springs in the eastern Pacific. In the way the swarm seems to hover in place, I was reminded of those clouds of gnats sometimes seen along the trail within a summer forest. At the vents, warm water flowing out of cracks in the seafloor rises up and then bends over in the current within less than a meter of the bottom. The amphipods swim into this flow of warm water with their strong swimmerets, but when they sense the upstream boundary of the hydrothermal flow they stop swimming and drift back-ward. A meter or so downstream, they start plying their way back to the source. It is as if they are on a hydrother-mal treadmill. Thousands of individuals, each not quite as big as a housefly, make up the swarm. These amphipods must rank as among the most active of deep-sea crus-taceans. I think the amphipods swarm not so much as a

Amphipod swarm
at vent orifice

social behavior, but simply because the treadmill kind of swimming strategy keeps them in the area with the most food. But I am only guessing when I think that their diet comprises microscopic bacteria suspended in the outflowing warm water.

Geologists seem to stumble on some of the really oddball animals, assigning them delightful descriptors. Spaghetti worms—long, white and thin, just as the name implies—drape in stringy heaps over pillows of basalt. It is impossible to tell from photographs where the tail of one worm ends and the head of another begins, they are tangled so. Technically, spaghetti worms are enteropneusts, a group of worms that normally live buried lives in soft muds. At Galapagos vents like Rose Garden, these enteropneusts are naked and exposed, and found mainly at the outside boundaries of a field. It seems such a vulnerable habit that I can only think the spaghetti worms must give off an offensive odor or that they are otherwise noxious or vile to the taste. Some trait must provide them with immunity from predation. Although I am often willing to taste deep-sea animals, the impunity of the naked enteropneust impresses me and I do not think I would eat a spaghetti worm voluntarily.

Geologist's dandelions are delicate golden orbs magically suspended in place just above the seafloor. A closer look reveals a galaxy of gossamer threads that reach out to anchor the organism to rocks or other surfaces. The orbs themselves are actually colonies of individuals that

Gossamer threads hold the dandelions suspended just above the basalt.

together make up spheres the size and color of a peach. Dandelions are siphonophores, related to the jellyfish-like Portuguese man-of-wars. If they are like their relatives, the dandelion's defense is a subdermal cache of stinging barbs. In clusters of five or six or more, dandelions—tangerine-sized—often tenant areas of dead or dying animals at vent sites in the eastern Pacific. I do not know if they are cause or consequence of the mortality.

The evolution and adaptations of the unique faunas associated with hydrothermal systems continue to raise questions. If we are to appreciate and exploit the variety of life on our planet, we need to understand the physiological and ecological attributes that allow these organisms to thrive in environments with some of the steepest gradients of temperature and in waters with compounds ordinarily toxic to life.

Over the past decade of exploration, we have learned that there are regional differences in the composition of species that make up vent communities. In much the same

Spaghetti worms draped over pillows of basalt at far edge of vent field

way that vegetation changes as one drives from Portland, Maine, to Miami, Florida, the fauna of vent communities changes along stretches of the mid-ocean ridge. Geographically distant sites are characterized by very different faunas. Whereas tubeworms, clams, and mussels dominate vent sites at spreading centers off the western coast of Mexico, a shrimp species dominates the fauna of known vent sites on the Mid-Atlantic Ridge. I do not doubt that important new types of faunal communities are waiting to be discovered in unexplored regions of the mid-ocean ridge system.

5

A CHORUS OF
TUBEWORMS

HYDROTHERMAL VENTS ON THE SEAFLOOR were not discovered by accident, although it is true that no one really anticipated the wealth of life to be found there. Expectations of seafloor hydrothermal venting were linked to concepts of seafloor spreading and plate tectonics, and so to the idea of continental drift. When oceanography came into its own in the postwar, Sputnik era, the first maps of the global ocean floor showed that mid-ocean ridges stitch their way through the basins. A Princeton scholar, Harry Hess, suggested that ocean crust is created at ridge axes as magma is forced up from the mantle; that same crust is eventually consumed in trenches along the periphery of ocean basins—he called this scenario "geopoetry." By imposing a birth and death

cycle for ocean crust, Hess's poem could explain why sediments older than the Jurassic are never recovered from
the seabed. This idea became known as seafloor spreading, and hard geophysical evidence to support the concept
began accumulating in the early 1960s. Once viewed as a
relic, continental drift and seafloor spreading evolved into
the modern concept of plate tectonics. It pleases me that I
am a contemporary of a concept in earth science that is as
fundamental as the notion that the earth is round or that
our planet revolves about the sun.

The theory of seafloor spreading may seem abstract,
remote. The time frame is geologic, the pace excruciating-
ly slow. But there is one place where seafloor spreading
takes on a most immediate sense: on the seafloor itself.
Seafloor spreading becomes tangible in the rift valleys of
the mid-ocean ridges. In *Alvin*, on the East Pacific Rise
and the Juan de Fuca Ridge, I have picked up pieces of the
very youngest bit of ocean crust on our planet—young
measured not in geologic time but in days or weeks. You
can see where the lava has welled up and spilled out onto
the seafloor. And you can drive along the newly formed
fissures that are the stretch marks of the birth of the
seabed.

As soon as the true nature of mid-ocean ridges was
understood, the more perspicacious among seafloor geologists expected that hydrothermal systems would occur
there. All of the necessary ingredients to maintain
hydrothermal activity exist: a source of heat from volcan-

ism, seawater as a fluid, and earthquakes that could crack the seafloor and make it permeable, so that seawater can reach the hot rock. Hydrothermal springs are common in volcanic systems on land; there was every reason to expect they would be common at the volcanic systems of mid-ocean ridges.

An expedition of geologists and geochemists took off with *Alvin* in 1977 for the Galapagos Spreading Center— a position in open water about 300 kilometers northwest of the Galapagos Islands—to hunt for hot springs on the seafloor. Their target was a specific spot where, earlier, a towed temperature probe had measured a spike of warm water. The signal they were chasing was subtle, less than a degree change in the temperature record over a very short distance, just a few meters. Yet the very first dive discovered warm water springs.

Two years later, biologists discovered the Rose Garden vent field where a riot of tubeworms stretched above the sponson of *Alvin,* well over 6 feet. I know this because I have in my slide collection one of the rare self-portraits of *Alvin,* taken by a National Geographic camera that *Alvin* carefully positioned on the seafloor. Moving away from the camera, *Alvin* posed beside Rose Garden tubeworms, providing unarguable proof of dimension. Among the tubeworms were mussels and clams and other invertebrates, but tubeworms distinguished the place.

A painted tropical fish can be just a lost rainbow among the gaudy carnival of colors of a coral reef; a vividly feathered tropical bird is veiled in a green lace of leaves and twigs. But giant tubeworms are 6-foot-long expletives, shouts of brilliance, startling in their vivid simplicity and exposure. Crimson plumes bloom atop long white tubes that emerge from cracks in glossy black lava. Warm water rising up from a vent ruffles the leaflike lamellae of

the plume. There must be a special cue in the vent efflu-
ent, a chemical billboard, that causes a cohort of tiny
tubeworm larvae to settle out of the water to grow together side by side as dense thickets of tubes because often the
tubes are nearly all the same size and growing in parallel
array. The growth of the whole guides the growth of one:
a chorus of tubeworms.

There is a singular place on the East Pacific Rise, in
the Venture Hydrothermal Fields, where I have driven the
submersible along the surreal landscape at the base of the
jagged edge of the shallow axial valley and seen looming 8
meters tall in front of me a tower of tubeworms. The
worms are so tightly packed that the basalt pillar they colonize is invisible. They look cultivated, each worm a prize
specimen arranged in a formal garden just so to present a
uniform face of red plumes about the pole. Because it is so
unique, this pillar has become by informal consensus a
sanctuary where scientists may go to look but not touch.
It won't be long, a few years probably, before the warm
water that feeds the tubeworms of the pillar ceases to flow
and the magic will disappear. Maybe only a dozen people
will ever see it. Already a patch of worms has sloughed off
and fallen to the seafloor to die and be eaten by crabs.

Gathering live tubeworms is an art. To be of much
use to biologists, the worms in their tubes must be carefully plucked and stowed in insulated containers for the
long ascent to the surface. The trick is to take advantage
of the worms' behavior. They have quick reflexes and dis-

appear into their tubes when disturbed. It is a reflex born of predation by voracious crabs that nip at the tender tubeworm plumes with their claws. Once you have caused the worms to duck, it is a simple matter to grab the tops of the tubes with one of *Alvin*'s manipulators, place a cluster of worms in the box, and shut the lid.

Bear in mind, though, that the pilot might be nearly upside down working the manipulator from the starboard viewport. The spot where the scientist had lain curled, busily taking photographs and taking notes, is preempted by the pilot. Scrunched into the back "corner" of the sphere, the scientist is now snug up against one of the oxygen bottles, feet folded uncomfortably, awkwardly out of the way as much as possible, knees a shoulder rest for the pilot.

The starboard manipulator is operated by a series of switches located near the viewport. Each switch feels different and controls a different manipulator function. A good pilot makes it all look easy by playing the switches like a musical instrument, fingers moving nimbly from one switch to another while at the same time twisting the knob that controls the hydraulic pressure so that pistons and motors translate commands into dexterous and precise motion. The manipulator becomes an extension of the pilot's arm. An experienced pilot can pick up an egg without crushing it, an exercise in pride practiced on deck. Sometimes to get a sample you have to reposition the sub just a bit by working the joystick with your knee while you work the manipulator with both hands.

I wish that I could get out of the submersible, not to collect tubeworms, but simply to run my hand along their tubes and feel the gradient of warmth in the water surrounding them. Of course, the water would stink of sulfide so that I would wish just as strongly to be back inside the submersible.

There is nothing obviously glorious about a tubeworm on deck. Pulled out of its tube and drained of its color, the worm is decidedly flaccid and unattractive. The vibrancy of the live worm on the seafloor is lost; in the thin atmosphere of the sea surface the plume loses its vitality and pales. I used to keep in my office a specimen tubeworm, preserved beside its tube in a jar of formalin, but it received justifiably unflattering reviews from guests. I traded it to the Museum of Nature in Ottawa in return for a prototype of a much more appealing fake tubeworm made of latex and foam.

Below the plume of a naked tubeworm is a gray, collarlike ring of muscle that gives the worm its ability to slip its plume in and out of the tube. Below the collar is a long, gray sacklike body. The tube itself, made of tough, flexible protein, is good armor for the vulnerable flesh of the worm. The tube is almost impossible to tear—good defense against crabs—but cuts smoothly with sharp scissors. The inside of the tube is slick and shiny, like a Teflon surface, that helps the tubeworm draw in its plume with haste when threatened. By some means, the inside of the tube is kept clean, uncolonized by microorganisms that foul the outside surfaces.

If one is guided by a good zoologist, a careful look at the external anatomy of a tubeworm body will reveal an attribute that sets this particular kind of worm apart from all others. It is actually the absence of features that is so revealing: in the adult tubeworm there is no mouth, there is no digestive system; there is no gut, no anus. Invertebrate textbooks teach us to think of worms as tubes within tubes, the interior tube being the gut; but that is not so in vent tubeworms.

It is quite something to discover giant tubeworms clustered around warm water flowing from the seafloor. But to find that these worms have no obvious means of ingesting food—this is truly the stuff of science fiction. Captain Nemo's captive companion, Monsieur Arronax, in Jules Verne's *Twenty Thousand Leagues Leagues Under the Sea,* does prophesy well when he says that if there are any mysteries left on this planet, they will be found in a special environment in the remote deep sea.

The mystery of how the tubeworm gets its food was solved in part by Colleen Cavanaugh, who at the time was a first-year graduate student at Harvard University. With persistent prodding, she will relate to you her story of the moment when all her thoughts came together in the answer and she jumped up in the middle of a class to tell it. It is the best of science stories, a contemporary discovery, one that makes us hope we might be as brilliant just once in our science careers.

Colleen found that answer inside the flaccid sack of the tubeworm body. It is a bloody job, but if you take a pair of fine scissors and slit open the thin body wall of the sack, you will see an odd-looking organ that fills most of the volume. The organ is soft and delicate, of irregular shape, branching and finely divided and highlighted throughout by yellow flecks. It smells bad. The yellow is elemental sulfur; the organ is the trophosome. Colleen prepared bits of the organ, placed them under high magnification, and showed that the trophosome comprises tubeworm cells filled with grapelike clusters of tiny bacteria.

It is the bacteria that somehow provide nutrition to the host. But in the give and take of biology, the tubeworm has to supply raw materials to the bacteria. What the bacteria need is a supply of sulfide, oxygen, and carbon dioxide. It is the requirement for sulfide that keeps the tubeworms close by vent water, which is laden with hydrogen sulfide. The catch is that the tubeworm also needs oxygen, which is in limited supply in vent water, if it is present at all. There is plenty of oxygen in the surrounding seawater, though. Tubeworms grow so as to keep their plumes in the region where the vent water and seawater mix. The tubeworm plume, reddened by a rich supply of blood and constructed of many layers of thin wafers of tissue, is a very effective gill that takes up sulfide, oxygen, and carbon dioxide. The blood pigment, hemoglobin, carries these compounds to the trophosome and the bacteria.

The rest of the story is in the bacteria, which take the inorganic compounds—that same sulfide, oxygen, and carbon dioxide—and produce organic material that the tubeworm can use. The bacteria are like plants, except that instead of using sunlight for energy as in photosynthesis, the bacteria obtain their energy from the sulfide in the vent water through chemosynthesis. If light from our sun is ever extinguished to a point where photosynthesis cannot go on, tubeworms could continue to thrive, sucking up the compounds needed by their bacteria.

No one could guess even the most basic things about Rose Garden or any other vent community during the earliest years of exploration. We didn't know how common vent communities might be. Or, if vent communities existed elsewhere, we didn't know if the organisms would be the same. Would the community change over time? What might cause change? Would entirely new kinds of organisms come to dominate, or would change happen so slowly that we wouldn't see it for decades? All of the ecological rules were undefined. For those of us lucky enough to be involved in this research, it is like discovering life on another planet and having the privilege of being among the first to study that life.

I joined a science team in 1985 to revisit Rose Garden. It was on this cruise that I made my first dive to the seafloor. I am glad that my first dive was to Rose Garden. It is a special place, both because of its significance as the first site to be described by biologists in the still young history of the science of hydrothermal vent communities, and because it has become the "textbook" vent community against which all other communities are compared. I remember that I spent the night before my dive wide awake in anticipation. In the calm heat of that tropical night, I quietly looked out over the surface of the sea while inside my mind whooped and shouted and cartwheeled at the thought of actually diving to the seafloor.

When I arrived at Rose Garden, the flourishing colony of tubeworms had all but given way to mussels. Golden brown mussels covered the walls of the main fissure and were heaped in mounds over smaller cracks between lobes of lava. We documented the change, but the reason for the change wasn't clear. It may have been that the hydrothermal plumbing was altered, clogged by minerals deposited beneath the surface or reorganized by an earthquake, so that the flow of nutrients could no longer sustain the worms. From the work of physiologists and ecologists, we know that mussels can cope and even thrive under nutrient conditions where tubeworms will succumb. Like tubeworms, mussels are hosts to bacteria that live on chemicals, but in the mussels, these bacteria are found in large, fleshy gills. The digestive system of the mussel is reduced, but functional, which may give the mussels an important advantage over the tubeworms.

Mussels are surprisingly good climbers. They have a foot that can secrete a tough thread or byssus to a surface, which might be another mussel, a tubeworm, or basalt. The attachment point is secured by a leaflike plate of adhesive and release is from the foot end of the byssus. By

Tubeworms are nearly overgrown by mussels

cycles of anchoring, bootstrapping, detaching, and reat-
taching in a new spot the mussel moves about, leaving a
trail of abandoned byssal threads. This movement is
slower than a box turtle crossing a road, not a speed easily observed in real time; but overnight, between dives, I
have seen mussels placed in the bottom of a box climb
their way up 20 centimeters to crowd around the lid. In
contrast, the tubes of a tubeworm are fixed and the worms
never abandon their tubes. I wonder if Rose Garden
changed because mussels were able to colonize the site and
crawl about, choosing the best areas of flow and over-
whelming and outcompeting the tubeworms?

Clambake, a graveyard of dissolving clam shells
lying in cracks where warm water once flowed, lies a short
distance from Rose Garden. Flesh of the clams was long
ago eaten by crabs and other scavengers. At Clam Acres
on the East Pacific Rise, where giant live clams form white
necklaces about lumps of lava in the heart of the field,
empty shells in now cold water along the margins are tes-
timony to an inevitable slow march of death as conduits
clog. I have flown *Alvin* over another site where a fresh
lava flow was frozen in thin black drapes over a portion of
a mussel bed. Just centimeters away from the chilled margin of the flow were live mussels. Death of a vent might
come slowly as conduits choke with minerals, or faster as
earthquakes modify flow, or catastrophically as lava over-
runs a community. The ultimate fate of all vent sites is
burial, if not by an inferno of molten lavas, then by sedi-

CLAM
BAKE

Some shells are
broken, others show
signs of dissolution, w/holes where
shell was thinnest.

ments as the ridge crest inexorably separates, subsides, and accumulates over eons the incessant rain of particles that sink through the water column.

If vents were far apart and rare in time and space, the cycle of birth and death of hydrothermal activity would challenge the existence and evolution of any organism specialized for that environment. How do species that colonize vents cheat the inevitable local extinction? One has only to think of the curious strategies adopted by terrestrial organisms—the frog that broods its young in its mouth, the butterfly that begins its life as a hairy caterpillar, the weed that secures its progeny inside a Velcro pod—to know that vent organisms may have novel tactics for survival that we have not yet imagined.

6

ON BROKEN SPUR

A S THE SUBMERSIBLE IS HAULED INTO THE hangar, I feel a distinctive shudder—the ship is brought to a stop. We are dead in the water, heading into a light wind and surface current. On the starboard 01 deck, the boatswain and his crew are lined up with long poles, hooks, and lines, ready to snag a transponder that is floating within a meter of the hull of the ship. The chief mate watches the pick from the small flying bridge that is cantilevered out over the water. From there she has a clear view of the starboard side and can relay commands to the sailor at the wheel on the bridge to hold the ship steady beside the bobbing sphere.

The transponder was released acoustically just after *Alvin* left the bottom for the last time in this area. In the top lab, the release code goes out sounding like a demanding electric buzz, an acoustic shout that travels through the water and is heard by the transponder more than 3,000 meters deep. *Alvin* can hear it too, a grating, harsh sound, at odds with the melodic pings and echoes of the tracking system.

On receiving the proper release command, a current sent through a wire on the outside of the transponder burns the wire clear through, dropping the 80-pound steel weights and 100-meter tether that anchored the transponder to the bottom. The buoyant and reusable glass sphere sheathed in a yellow plastic hardhat and its enclosed electronics are freed and begin their ascent.

The *Alvin* pilot standing watch as surface controller in the top lab monitors the ascent of the transponder and the sub with acoustic tracking instruments and software, telling the bridge when and where both will surface with an accuracy that is good to seconds and meters. *Alvin* is recovered first, and then the ship is brought into position to pick up the transponder. This time the boatswain himself makes the snag, and the instant the transponder is brought on deck I feel the ship's engines put full ahead. We are off to the next dive site.

We head north into the night on a course that parallels the strike of the Mid-Atlantic Ridge invisible beneath us. It will take a night and a day and another night of steaming to reach our last dive area, the *AII* making a fair 10.5 knots over the water.

Our transit is a quiet time for the ship, a chance to catch up on repairs and paperwork, samples and conversation. We have already made fourteen dives in this series together with our French colleagues. They and their ship, the R/V *Jean Charcot,* had been our companions in the middle of this ocean, joining us on dives and sharing samples collected by *Alvin.* Our joint program is finished now and they are going home. The *Charcot* was a good neighbor. We will miss our colleagues and the companionable bright lights of the *Charcot* as she held station beside us through the nights. The courtesy French flag borne beneath the American flag on our mast is lowered.

I like the feel of the ship under way. On calm evenings like this one, we slice through the water, curls of waves shedding off the bow, collapsing endlessly one on another to either side. I look for the planktonic lanterns that sometimes light our way, but this night I do not see any bioluminescence. In the right place and time, water chasing off the bow is illuminated by a soft blue-green glow as millions of microscopic organisms flash in protest to the jostling water. Flying fish take off across the flat waters, gliding for such long moments that I wonder if they won't ever drop back into the sea. On the fantail, the wash of the stern props boils the sea and leaves behind a broad wake that stretches aft beside a path of moonlight. The bow waves, the wake, and the steady plot of our progress by the ship's satellite positioning system are the tangible evidence of our orderly progress over the sea.

California
Flying Fish

Cypselurus
californicus

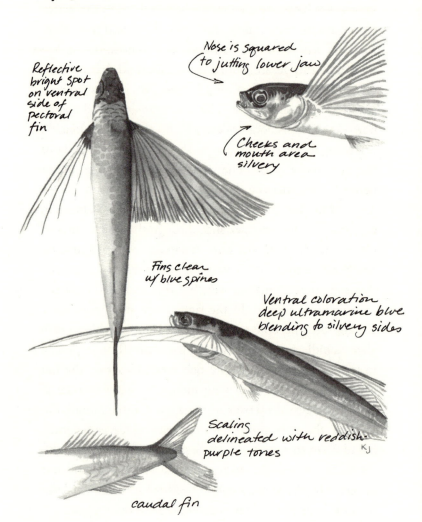

Nose is squared
to jutting lower jaw

Reflective
bright spot
on ventral
side of
pectoral
fin

Cheeks and
mouth area
silvery

Fins clear
w/ blue spines

Ventral coloration
deep ultramarine blue
blending to silvery sides

Scaling
delineated with reddish-
purple tones

KJ

caudal fin

Our next and last target is a hydrothermal vent site known as Broken Spur on the Mid-Atlantic Ridge. Quiet anticipation has been building during our cruise. Broken Spur, discovered by British scientists led by Bramley Murton, has never been visited by a submersible before. We do not know entirely what we will find.

Broken Spur is only the fourth vent field to be discovered on the Mid-Atlantic Ridge. The first vent sites, at TAG and Snake-Pit, were found in 1985. For several years after that, little scientific activity took place on the ridge crest in the Atlantic. Then, in the fall of 1992, Charlie Langmuir, from the Lamont-Doherty Earth Observatory, was chief scientist on a geological cruise that dredged up a chunk of sulfide from Lucky Strike. A few months later Broken Spur was located with a towed sensor package, and the number of known hydrothermal systems on the Mid-Atlantic Ridge doubled.

When I learned about the discovery of Broken Spur, I checked the latitude and longitude of the site against the dive sites I was preparing to work with *Alvin* in two months. I knew the answer even before I looked: Broken Spur at 29 degrees north was located between our port of departure, Ponta Delgada, Azores and our southernmost dive site, Snake-Pit, at 23 degrees north. We were planning a brief three-day stop at the TAG site at 26 degrees north as well. With *Alvin* right in the neighborhood, the chance to dive on an unexplored vent site was compelling. Our science party, the ship schedulers and the funding agencies all agreed and, with only a few weeks before our scheduled sailing date, two dives were added to our cruise.

During the final transit, Bramley Murton, a fair-headed Brit with a name that seems straight out of a Dickens novel, pulls out his charts and maps of where the new vents ought to be. Together with *Alvin* pilots Pat Hickey and Bob Grieve, we study and scheme, laying our best plans for the next day's dive. We know the terrain in the target area is complicated, rugged. We will need to use all the clues at hand if we are to find the site in our allotted two dives.

Now our first effort to find the site with *Alvin* is less than twenty-four hours away. Bramley targets the western slope of the ridge axis for the start of the dive, just north of his best estimate of the location where the camera sled picked up sulfide. We prepare the science basket at the front of *Alvin*. It soon bristles with tools that will allow us to do a little bit of everything if we chance on hot water and animals. We carry titanium samplers for the water; push cores for sediment; bins to hold pieces of sulfide; a closable, insulated box and a slurp gun for animals; and a temperature probe.

We finally arrive on station and are submerged by midmorning. It takes nearly two hours to reach the bottom at a depth of just over 3,000 meters. In the greenish cast of the thallium iodide driving light, Pat lands us gently on pillow basalts.

Only the surface of the moon seems as remote and

desolate a destination. We should be glad we skim over these fields of pillows in *Alvin*. Any traverse by foot across this kind of terrain would be exhausting and treacherous. No rhythmic gait would be possible, given the uneven throw of the basalt pillows, and spaces between the curves of the pillows are just large enough to be ankle twisters. Small fists of glass sprout from the pillows. I know from experience that these pillow buds are easily broken off—they make good samples for geologists and are easy to pluck with the manipulators. If they slough off as readily when stepped on, they would turn the concave pillow surfaces into slippery shingles of knife-like black glass shards. Soles of hiking boots would be cut to shreds.

To get our bearings, we drive west from our landing site and see that our depth increases slowly, consistently. We are out on the flank, driving away from the ridge axis, away from the neovolcanic zone, away from our target. Moving back to the east, we reach the shallowest point, 3,050 meters. Since we want to remain on the western flank, we switch to a roughly north–south search pattern, a 10- to 15-minute run in each direction, with a short slip to the west on the turnarounds.

Striated pillows continue before us as we slowly slalom to the west. Fault scarps and fissures scar the seafloor. My eyes and head begin to ache from looking so intently for any sign of hydrothermal activity in the area. I want to see a bit of fresher-looking lavas, maybe some

orange or green or red staining, a dusting of glittery sulfides, an increasing number of organisms, a certain murk-
iness to the water. But all we see are pillow lavas, varying
minutely with fewer or more buds, perhaps larger or
smaller lobes. In my audio record of the dive, I run out of
adjectives to describe the subtle variations in the lava morphologies. It is a relief to the eye to glimpse a solitary
crinoid or perhaps a passing fish. Once we pass a gigantic
sponge, floppy lobes of white sponge of a sort I'd never
seen before. I grab my camera and shoot off three frames.

We are prudent and deviate little from our survey
plan, waiting for a real sign that we are close to a vent. For
a brief stretch we pass over lavas that form flat lobes
instead of pillows. This seems a good indication, but we
are over the flow for only a few minutes and we spot nothing. At one point we recognize we have circled back to our
landing site.

By midafternoon the surface controller topside has
finished the survey work and is tracking us. We are near
our intended target and head directly there with a vector
supplied from above, but we find nothing. We return to
our north–south sweeps.

Two hours into our bottom time, the sonar indicates
mounds up ahead. We maintain heading and soon Bramley makes out red, oxidized sediments. I see increasing
numbers of anemones and a couple of empty mussel
shells. Sulfide deposits, old and weathered, cold and inac-

tive, show up on the video camera that is looking forward, sharing the same view as the pilot.

Rusty sediments pond in shallow depressions between the weathered gray curves of basalt pillows. As the submersible glides over them, clouds of orange dust rise, swirl about, and slowly drift back down. The fine sediments are fallout from a hydrothermal chimney that had been belching out smoke for years. When the particles first rain down, they are black and glittery, finer than the finest soot, but as they accumulate in pockets on the seabed, they rust to orange iron oxides. Some kind of animal finds it a suitable habitat and has colonized the sediment: flared, trumpet-shaped tubes poke out of the mud. I watch closely to see who lives inside the tubes, but my patience is unrewarded. Nothing emerges. We are able to sample a few of the tubes by vacuuming the sediment. Later, on deck with the tubes placed under a microscope, I will find the small segmented body of a polychaete worm. Long palplike tentacles adorn its head; they must emerge from the tube to gather food when the worm is left undisturbed.

We are sure now that active sulfides must be nearby, so we abandon the relic site after a few minutes of reconnaissance to chase down another sonar target up ahead. Four minutes later, after a brief transit over pillows spotted with small white anemones, we arrive at another low mound. Warm water flows out from beneath the edges of irregular lobes of metal sulfides. As it rises, the warm water mixes with the cold ambient seawater and shimmers. Pat focuses the color camera on the flow, and I make a quick sketch of the mound in my notebook. There are animals, but in contrast to other vent systems I have visited, the biology here is depauperate. Shrimp, anemones, and brittle stars dominate, but their numbers are few, their biomass small. Tubes of another kind of polychaete worm lie in shallow ripples on the sulfide surface and look like a few scattered straws of a child's pick-up sticks. I have seen these kinds of animals elsewhere, in other vent systems. The brittle stars are perhaps the most interesting, because little, if anything, has been published about their occurrence at vents.

We sense the beginnings of a puzzle: Why doesn't this water support more animals? Is it depleted in the nutrients that chemosynthetic organisms require? Is this a dying site, or is it just in the early stages of reactivation? Will we find other vents in the area that have higher levels of biological production? Could there be an unseen marauding predator that keeps populations of other animals in check? Or are the few crabs that we see so voracious that any new recruit stands little chance of survival?

We decide not to sample here, but we do leave a small marker before taking off again, heading south. Within a few minutes we approach a pile of broken-up plates of sulfide. Five meters farther is another low mound, this time with a chimney rising out of it. Pat drives up to it and begins a vertical ascent.

Bramley and I watch the chimney on the video. Incredibly, the spire thins until its diameter is less than 20 centimeters, thickens again, then narrows and leans crazily to one side. It seems improbable that it is still standing. We continue the ascent and finally reach the top, where the spire spreads out once more, top-heavy at 20 meters above the seafloor. More than a dozen jets of hot water crown the apex and create a billowing cloud of black smoke that rises above us. The jets are arranged like stout stiff fingers of gloves, clustered in short rows; the entire top meter of the spire is leaky with wispy sheets of smoke streaming out of a crenellated, shingled surface.

Pat and I both know this spire must be fragile and will topple with the slightest nudge from *Alvin*. Gingerly, he reaches out with the manipulator and plucks a piece of sulfide from the growing edge of a jet. A few shrimp tucked in a crevice become the target for the suction hose. Shrimp slip into the sample chamber just as the spire tumbles over. We maneuver about and relocate the remains of the spire. There are still 12 meters of chimney left, and now the hot water gushes out of one or two openings. I suspect the spire has toppled over many times on its own, perhaps shaken by an earthquake or just from its own instability. I am sure that it will grow back quickly.

With only a few minutes left in the dive, Pat measures the temperature of the hot water—the probe reads a steady 354°C when held in the throat of the largest orifice. We ask for and get permission from the surface to stay long enough to get a hot water sample. That task accomplished, and with the pinger and a marker deployed so we can return to this site tomorrow, Pat calls the surface and lets them know we are ready to come up. Cleared by the surface controller, he moves away from the spire and drops his weights, and we leave the bottom.

As we ascend, Bramley and I still don't know what to make of the paucity of life here. Even the spire had few animals, at most a few dozen shrimp and some brachyuran crabs. In contrast, at TAG and Snake-Pit to the south, shrimp cover the entire surface of the sulfides in swarming masses, shrimp bodies packed so tight as to be touching. Despite our success in locating the site and the solid knowledge we gain from discovering what lives there, I can't help but feel disappointed that the new vent field is not home to some fantastically new kind of sea beast.

7

BLACK SMOKERS

G IVE ME A DIVE WHERE I AM FREE FROM ANY obligation to collect samples or data, a day just to do whatever I want as if I were on a picnic on a lazy Sunday afternoon. I will spend that day in a field of black smokers, just looking.

Raw and powerful, black smokers look like cautionary totems of an inhospitable planet. Like the undersea volcanoes that drive them, black smokers are born of primal forces, a consequence of first-order geophysical processes that control the motions of our ocean's crust. I have often worked black smokers in *Alvin* and I never fail to be awed by them. Approach the simplest structures: they first loom as tall dark shadows, black against surrounding blackness at the gloomy edge of *Alvin*'s pool of

light. Maneuver closer. Illuminated now, the base of the column rises up along the side of a shallow fissure. Small pieces of sulfide rubble, failed fragments of the larger structure, litter the seafloor. A fine dusting of sulfide covers the nearby flat surfaces of sheeted lavas. The stack itself is roughly textured, with irregular walls. As you ascend the chimney, watch for hot water. It is usually at the very top, sometimes 6 or more meters above the seafloor. Chimney walls may be leaky and alive with animals drinking up a warm seepage of noxious chemicals, or the chimney might be hot and young and sterile, devoid of life.

At the top of the stack the challenge for the pilot begins. You have to get the sub stable against this pillar so that you can get to the orifice to sample and make measurements in the hot water. There is an art to setting up. A pilot can make *Alvin* hover, neutrally buoyant. To work the top of a chimney, you get neutral, and then you position the front of the basket or the aluminum ski that sticks out beyond the basket so that it rests like a bumper against the structure you want to work. Put *Alvin* into autopilot to maintain a constant heading; go ahead a little on the pot that controls the forward thrusters. Maintain a little bit of constant pressure by driving against the chimney wall so that you stay still in one place. You have to trade off stability with power consumption. Drive too hard against the chimney and you will use up your batteries too fast; don't drive hard enough and you will not have control of the sub.

At first, once you're set up, you hold your breath and give stern looks to anyone who even thinks about moving suddenly in the sphere. Sometimes the stack just isn't strong enough; it topples over. Damn. But usually you get steady pretty easily and can settle down to work, the black smoke just in front of you, within easy reach of the manipulators.

Only once did I abandon an effort to sample a specific black smoker because I could not safely get into position. It was at the end of a long dive, at a site I think of as Inferno, but which is mapped as Dante, a very large and complex sulfide structure on the Endeavour Segment of the Juan de Fuca Ridge. More a tall mound than a chimney, Dante has hot water blasting out from small stacks that project from the larger structure. Two other large active mounds, Lobo and Grotto, form with Dante a U-shaped system of sulfides within an area about half the size of a football field. The targeted black smoker orifice was inside that U, where there was just enough room to get in with *Alvin*. To get out, my only option was vertical, up. Black smoke made visibility poor. I was neutral and at the mercy of a confusing current—upwelling and side-drafting—set up by the vigorous venting and complicated by the confined geometry of the site. I could not get to a stable position without jeopardizing some part of my sub, and had to give up.

I have seen the side of the sub after a pilot unknowingly brushes too close to a smoker. It is not pretty. The painted white fiberglass skin gets covered with black soot, fire-damaged. Where the hottest water blasts against the sub, the paint and fiberglass blacken, blister, and melt. Once, when control of one of the manipulators failed, plastic tubing that carries hydraulic fluid to the manipulator wrist and hand was cauterized as it became immersed in hot water, making that manipulator immobile. Scientists around the world display on their shelves, like trophies, pieces of sampling gear impossibly scorched by water.

Being in *Alvin* isolates us from anything but a visual impression of a black smoker. So we rely on instruments and probes to provide vital information. The first time *Alvin* approached a black smoker, no one was quite sure how hot the water was, but the melting of a plastic milk crate inadvertently stuck into a jet of a black smoker provided a minimum measure of the temperature of the hot fluids. Now we routinely carry temperature probes on *Alvin*. Each is made of a slender cylindrical pressure housing for the electronics, and cables for power and data emerging from one end of the cylinder to connect to the sub's electrical penetrators, with a long titanium stinger projecting from the other end. One of these probes sits permanently just beneath the viewport on the bottom of the sphere as a safety measure. Pilots monitor this probe when we work in hydrothermal fields; we will abandon a site with impressive haste if the temperature read by this probe starts to rise too much above a few degrees Celsius.

The temperature field around a black smoker always surprises me. When I first learned of black smokers, I had expected that a large volume of water would be heated to dangerously high temperatures, but this is not the case. If you pick up a temperature probe with a manipulator and position the tip of the stinger just a few centimeters away from the edge of the strongest outpouring of vent fluid at a black smoker, the temperature will read a frigid 2°C. Now creep the probe closer until the tip almost but not quite touches the jet of water . . . still 2°C. Touch the jet and the temperature jumps to 350°C or more. Over a distance of a few millimeters or less there is a 350°C gradient of temperature. I am sure this must be the steepest natural gradient of temperature on the surface of our planet. Rising hot water pulls with it a surrounding sheath of cold water, effectively insulating the base of the jet. If you move the probe vertically above the jet, the temperature also falls off precipitously, returning to near ambient bottom water temperatures within only 10 to 20 centimeters. At the orifice, the power of the jet is visible, the vent fluid flowing out with the force of a fire hose. It looks as if it should roar, but there is no sound that can be heard through the hull of the submersible. The jet becomes turbulent as it moves upward. Small eddies in the flow entrain cold water, mixing it into the plume, cooling the plume quickly.

It is hard to conceive of how hot 350°C really is, but pilots have a healthy respect for these temperatures since, like that plastic milk crate, the acrylic viewports of *Alvin* melt at ~80°C, a temperature well below that of a black smoker. In Fahrenheit, temperatures of black smokers sound even more impressive, greater than 600°, hotter than molten lead.

It is the steep temperature gradient that makes it possible for us to work black smokers with a large measure of safety. A pilot quickly learns the region where the hazard lies. We can, carefully, put the front of the sub within a meter above a black smoker orifice. If you have the sub ballasted to neutral when you do this, you can ride the thermal plume vertically—riding the elevator, we call it— taking samples of the plume water as you ascend. As you ride upward, you become enveloped by the billowing cloud of black smoke unless you maneuver to stay right at the edge of the plume. Challenged by geochemists, I have followed the plume as high as I could by keeping the smoke in sight. I traced it up to nearly 80 meters above the orifice before the plume became too diffuse and dilute for me to see it. Using sensitive temperature probes that detect thousandths of a degree differences with accuracy and transmissometers that measure the amount of particles in the water, we know that the plumes rise upward to as much as 200 meters above the seafloor before they become neutral relative to the surrounding water and begin to move laterally.

Black smoker fluids are not just hot. They are acid solutions, depleted in dissolved oxygen and magnesium, and laden with minerals stripped from rocks deep within the ocean crust. Vent fluids carry a burden of chemicals that include dissolved iron, copper, zinc, and other metals and volatile gases like hydrogen sulfide and methane and carbon dioxide. We are cautioned as pilots not to take on ballasting water when we are in a position where diluted vent fluids might be near the intake, for fear of damaging our titanium storage spheres with caustic water.

There are different kinds of smoker structures in different places. On the Mid-Atlantic Ridge, at the vent field known as Snake-Pit, there is a site called Moose. At Moose stands a single large sulfide structure that is nothing at all like a simple column or stack. There is no mistaking Moose. Sulfide outgrowths that look undeniably like Moose antlers project out from the main mound. A few small jets of hot water flow from stubby projections among the antlers. Beehive, another sulfide edifice, is a five-minute submersible ride due west of Moose. Although it is so close to Moose, it lacks any mooselike attributes. Instead it has numerous beehive-shaped patches of shingled sulfides where high-temperature water streams upward.

There is Angel Rock, a fairyland of low white anhydrite spires and shelves in Guaymas Basin; Pipe Organ, an array of fused tall black tubes on the Juan de Fuca Ridge. Among pilots, one of the most infamous of black smokers is Godzilla, a sulfide structure on the Juan de Fuca Ridge that is a stunning 45 meters tall (taller than a 13-story

Beehive is shimmering

Wisps of black smoke escape from the edges

Sulfide "sculptures" burnt sulphur yellow colors

"*Beehive*" and "*Moose Antlers*"

building), 12 meters in diameter, and still growing. There is an artist's drawing of Godzilla based on a mosaic of photographs analyzed by scientists at the University of Washington; it is far too big a structure to be imaged in its entirety. *Alvin* drawn to scale is dwarfed by the giant. Godzilla is ornamented with large sulfide flanges—

irregular shelves of sulfide that grow outward from the main trunk of the edifice, like giant shelf mushrooms on a giant tree. Hot water pools beneath these flanges, creating unique hazards for a pilot ascending the structure. Although I have seen flanges in several places, nowhere are they more spectacular than on the Juan de Fuca Ridge.

When you work a flange, you park the sub right in front of it. If you position the sub just a little beneath the flange pool, you can look up at it. Because of the density difference between the cold seawater and the hot water, the pools are as reflective as a mirror. It is disorienting at first. Using the video camera mounted on the manipulator, you can look straight up into the pool and see through the water to the delicate fingers of sulfide that are growing downward. It is a magical, upside-down world. The top of a flange is covered with biology—limpets and tubeworms and polychaetes. The animals pile one on top of another in a confusion of life. If you could get out of the submersible, you could scoop the animals up by the handful.

Black smokers don't last forever. In death, when the flow of hot fluids is cut off, sulfide structures turn orange with rust—iron sulfides turned to iron oxides. The vent animals die and decay or are eaten by scavengers. I have learned that, like a dead tree, a tall inactive sulfide structure can chemically rot to a fragile shell through oxidation, ready to collapse at the first jar of an earthquake or, to my dismay, a tap from *Alvin*. In an instant, several hundred pounds of rock from a decomposing chimney once tum-

Wisps of black smoke overflow from "puddles" under flanges, and from smoking vents

Tubeworms adorn the top of the sulfide structure

Flange fauna

Hot water pools up under flange

bled into my basket, carrying the sub swiftly to the seafloor. I was able to control the descent with my thrusters to keep us from getting snagged in a fissure just beneath us and to soften our landing. Then I spent a

Brisingid seastars face into current on inactive sulfide chimneys.

tedious half hour removing bits of punky, rotten chimney from the basket with the manipulators, piece by piece, so we would be light enough to leave the bottom.

Sometimes dead sulfide structures maintain their integrity long enough to be colonized by a new set of animals. The structures provide a measure of vertical relief sought by deep-sea invertebrates that make their living off particles in the water column. Wherever there is an inactive spire, three or four orange brisingid seastars are likely to cluster near the top. Their long spiny arms form filigree cups that face into the current, passively ready to capture bits of food that float past.

Less than twenty years ago, no one imagined that black smokers existed deep below the surface of the ocean. I know I am lucky to be as familiar with them as I am now, to know something of their power and drama, their variety and distribution, and to be certain that there is still far more to be learned about them.

8

CHASING ERUPTIONS

T HE FIRST SOUNDS WERE LIKE A HUMMING— the strumming sound of lava in motion through crust. Over several days, swarms of earthquake tremors propagated along the strike of the ridge axis, marching along a trend just east of north, 020 degrees. By the time the event was over, less than three weeks later, it had covered some 30 kilometers. Thousands of meters beneath the sea's surface, the earth had squeezed up a shield of molten lava—a dike—that moved north-north-east through a fracture beneath the seafloor until it reached a weak spot where it could break through to flow over the seafloor. We heard it and, incredibly, we knew we had heard it. A volcanic eruption on a segment of the Juan de Fuca Ridge, off the coast of Vancouver, was heard on the navy's SOSUS system in June of 1993. (SOSUS is a Darth Vadar–like antisubmarine SOund SUrveillance System of moored hydrophones off the northwest coast of the United States.)

Eruptions are apocalyptic events on the seafloor. I came close upon one when I was a pilot working the East Pacific Rise in 1991. Fresh lavas and bacteria prevailed along with pervasive venting of warm waters along the long, linear fissure from which the lava erupted. At the site called Hole-to-Hell, black smoke billowed up in a dense cloud directly from still-cooling basalt on the seafloor instead of from sulfide chimneys as we are used to seeing. Hot water poured out the tops of basalt pillars that normally stand as cold obelisks in the middle of drained-back lava ponds. I piloted the sub over a ridge axis covered with thick drifts of bacterial matter (called "floc") and mats. Where the floc blew out from skylights in subsurface caverns, they created a blizzard, a submarine whiteout, and I was forced to drive the sub by sonar. At the Barbecue site, scorched tubeworms were scattered over the seabed, tubes blackened, flesh intact but charred by a putative explosion. Scavenging crabs move in to feed on dead tubeworms. Tubes of worms we bring back to the surface smell of decay, the flesh inside is liquified, repulsive. We looked in vain for vent communities mapped earlier. But where the lava erupted, whole communities have been vaporized. Until now, the ocean had hidden this violence from us.

For years I—we—had wanted to catch a seafloor eruption in the act. Most of the ongoing volcanic activity on this planet—more than 80 percent of it—takes place along the submarine mountain ranges that make up the mid-ocean ridge system. Events like the one recorded by SOSUS may be the fundamental unit—the marine geologist's quantum—which, occurring repeatedly through time and space, accounts for the creation of the ocean crust. Who wouldn't want to rush out and take a look?

When the call comes from John Delaney and Bob Embley to join the cruise on short notice, I do not hesitate. I pack my gear and fly to Seattle to meet briefly with Delaney and his colleagues at the University of Washington, and then head down to the ship docked at the Port of Astoria in Oregon, near the mouth of the Columbia River. With two geophysicists, Paul Johnson and Maurice Tivey, in the car, I take advantage of my captive audience, asking questions about crustal permeability, porosity, tortuosity, magnetics, gravimetry. I receive a short course in geophysics, punctuated by the tectonic lessons of the region through which we are driving. I learn that as much as 20 percent of the volume of fresh lavas is water-filled voids. Not surprisingly, as lavas age and weather, their porosity decreases.

As we drive on, the Willapa Hills of coastal Washington come into view. I learn that, like the great ore-bearing Troodos massif of Cyprus, the Willapas are bits of seafloor crust that have been uplifted and reorganized to become part of the subaerial domain—ophiolites. The entire Olympic Peninsula is an ophiolite. Evidence of past hydrothermal activity, when the Olympic Peninsula was submerged, is found today in areas of hydrothermally altered rocks.

Over coffee at a truck stop in the logging town of Aberdeen, I learn that fresh, solidified lavas are more magnetic than old lavas and that this magnetic anomaly decays over time. But above what is known as the Curie temperature, lavas are nonmagnetic. Tivey and Johnson think that, at the site we are going to, some volume of the fresh lava flow beneath the surficial chilled crust may still be hot and demagnetized.

Before we sail, the science party assembles for a meeting. Among us are geologists, geochemists, microbi-ologists, ecologists. Delaney and Embley lay out our strat-egy. The first dives will be in the northern region, at the Flow Site, where the lavas escaped onto the seafloor. A second target is an area a few kilometers to the south known as the Floc Site, named for the chunks of white flocculent material, presumed to be of bacterial origin, blown out of the seafloor from warm water vents. Delaney thinks that, while the area of fresh lavas is interesting, the roots of the volcanic activity must lie to the south, where the earthquakes first originated. He predicts that if we are to find high-temperature vents, they will not be where the lava broke through the crust, but back at the subsurface origin of the lava, what he calls the Source Site.

We go no further than mapping out the first few dives of the series, knowing that the science will evolve in directions we cannot predict as we begin the expedition. With the immediate logistics under control, Delaney queries the party for a consensus on the name of the area where we will be working. We are on a part of the Juan de Fuca Ridge that hosts the Axial Seamount, inspiring National Oceanic and Atmospheric Administration (NOAA) scientists on an earlier cruise to the eruptive site to refer to it as the CoAxial area. The name sticks.

The first dive finds the new lava easily and reports back that the temperature of the venting water has decreased. The second dive I share with a geologist. We are to start at the southern tip of the lava flow and work our way down along the axial fissure to the Floc Site. We land on fresh lava and first drive east to the contact between new and old lava. The contact is unmistakable: new lava, glossy black with occasional orange staining between large pillow mounds contrasts with the smaller gray pillows of the older seafloor. We follow the contact south, our dive track outlining the edge of the flow. I note a solitary yellow crinoid perched on a small island of old lava, surrounded by fresh lava, unperturbed by the inferno that flowed within centimeters of its stalk.

We find the southern terminus of the flow, continue south along a fissure, and work our first sampling station at an area of warm water that vents at the base of the western wall of the fissure. The pilot measures tempera-

contact between old & new lava

tures and samples water and rock and an orange microbial fluff. Farther on we find a second area of warm water and repeat the sampling suite. It is difficult to tell the age of these low-temperature vents. Did the recent eruption trigger the flow of warm water?

Toward the end of the cruise, weather keeps us from diving. We stand by, waiting for the front to pass by or for our allotted time at sea to run out. It is not clear which will happen first. We are an idle ship in a stormy ocean. The mates keep us on the most comfortable course, heading upwind and then turning smartly 180 degrees to run before the wind, quartering the seas. *Alvin* is buckled to the deck with retaining gear, and the fantail, awash with the sea, is secured, the watertight doors sealed and posted NO ACCESS. When the ship rolls on the first turn, we discover the bits of science gear that did not get tied down with rope or bungee or duct tape. Inevitably a turn is made at mealtime, sending galleyware crashing onto the deck and causing a great deal of commotion and comedy at the mess tables.

Finally the seas subside and *Alvin* goes in the water again, with Delaney on board for the last dive. For two hours the sub encounters nothing but a Nike shoe among the pillow basalt. Around lunchtime, the surface controller monitoring the dive gives them a vector that takes them straight to the Source Site, where they find sulfide relics and active black smokers.

For the most part, the sulfides of the Source look to be far older than the two months that have elapsed since the eruption, but there is a fresh-looking anhydrite chimney sprouting from the side of one of the sulfide features vigorously emitting hot water. The chimney could arguably have had a coeruptive origin. Delaney has the pilot measure the water temperature of the new chimney and finds it to be not exceptionally hot, only about 280°C. We had hoped to find 400°C vents, a certain sign of complicity between magmatic intrusion and hydrothermal plumbing.

The anhydrite chimney is uncolonized and crumbles to dust when the pilot works to collect a water sample. But the adjacent, older sulfide chimney is covered with a ragged beard of a slender species of tubeworm. Red knots of miniature palmlike tentacles mark the presence of alvinellid polychaetes. Something has enticed dozens of new recruits of a type of galtheid squat lobster to the area, each recruit no bigger than a quarter, a miniature replica of the adult. The tubes of the tubeworms may provide some clues about the history of venting here. For much of their length, the tubes are stained brown by manganese, but there is an abrupt transition to an unstained, lighter coloration near the growing ends. It seems likely that the nature of the hydrothermal environment changed at the time when the tubes stopped picking up the brown stain. We don't know yet whether this coincides with the eruptive event recorded by SOSUS.

Mature sponges and soft corals, not usual components of the vent community, colonize the margins of the vent environment at the base of the sulfide structures. The sponges don't look like sponges to me. A thin, pale orange tissue covers inch-long spines that form comb rows around a thin stalk. In some places, sponges and tubeworms intermingle, a circumstance that I would not have predicted likely. Although I had never seen it before, this sponge is reportedly not uncommon in the deep waters here. The soft corals are fleshy pink and flowerlike, growing in a low colony from a communal mat, the stubby coral polyps emerging from the mat in tight and uniform array. They look as although they belong in a warm and sunny tropical coral reef rather than the dark and cold abyss.

Biological sleuthing suggests that not long ago these structures were all but inactive, with just barely enough flow for the tubeworms to subsist, but not enough flow to exclude invasion by the nonvent suspension feeders—the corals and sponges. Did the eruption revitalize the site?

Where the seafloor erupted on the East Pacific Rise, biologists and geologists have paved the seafloor with plastic benchmarks in linear order. Each marker is placed within sight of the next, and each is uniquely numbered. Now *Alvin* can fly the path repeatedly, documenting how vents along the kilometer and a half of fresh lava change over time. In the first year, the venting consolidates. One year after the eruption, Hole-to-Hell supports a collection of active black smokers grown to several meters in height. Elsewhere along the annotated submarine highway, patchy, persistent low-temperature flows are observed. The bloom of bacteria has dissipated, but bacterial mat still covers basalt surfaces where the warm water fluids percolate through the crust.

Crabs are among the first animals to colonize the new vent sites. They surprise us with their voracious consumption of bacteria. In monotonous cadence, the feath-

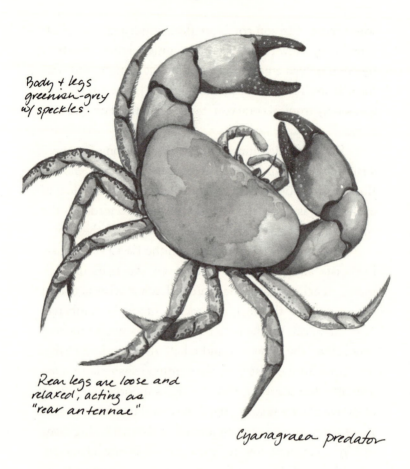

Body + legs greenish-grey w/ speckles.

Rear legs are loose and relaxed, acting as "rear antennae"

Cyanagraea predator

ered tips of maxillipeds that frame their mouths reach out to sweep over the mats. We guess that some of the crabs have wandered into the new vent fields from nearby areas unaffected by the eruption, the arthropod assault facilitated almost certainly by a sharp gradient of chemical signals. But I recall spotting hundreds of juvenile crabs as I drove over parts of the caldera just after the eruption— leggy white crablets scattered over a field of fresh black lava. I wonder if those juveniles were not the advancing front of the crab invasion.

The Jericho worm is also among the first of the more conspicuous colonizers. Small Jerichos plaster patches of basalt in the new flows of warm water. Jerichos are tube-worms, differing from the giant tubeworm by their small-er size and especially by their concentrically pleated tubes. Mature clumps of Jerichos, tubes twisting serpentinely about each other, look like Medusa's head.

The fecundity of adult Jerichos must be legend, the navigation abilities of the larvae extraordinary. We know from our observations that at least one cohort of larvae quickly found the new hot springs, but exactly where the larvae came from and how they located the vents are unknown. Are they offspring of nearby populations of Jerichos that spawn in response to the pulse of chemicals released during the eruption? Or do the new recruits rep-resent a bolus of offspring from an assortment of brood stocks from nearby and farther upstream that by chance passed through the area at an opportune time? Are they always among the first colonizers, the weeds of this hot spring ecosystem?

For biologists, eruptions on the seafloor are the abyssal equivalent to the emergence of Surtsey off the southern coast of Iceland. Surtsey was born of the

Mid-Atlantic Ridge and was quickly set upon by scientists keen to follow the natural history of an oceanic island. Now we have the chance to follow the biotic, geologic, and chemical history of hydrothermal vents from birth to death by periodically returning to eruption sites and documenting changes.

9

"BLIND" SHRIMP

"BUT DO THEY TURN PINK WHEN THEY'RE cooked?" I am asked, as I try to describe the gray-beige shrimp living at hot springs deep in the Atlantic Ocean. It is a reasonable enough question, since the shrimp crowd around plumes of black, 350°C water pouring out of sulfide chimneys on the seafloor. The shrimp are protected from the cauldron, though, by seawater drawn up beside the rising plume. Further, the heat escaping from the earth's interior is quickly absorbed by the surrounding seawater. Within a few centimeters above the orifice, the temperature of the plume is a comfortable 20°C and within a meter it is an icy 2°C. Still, wouldn't the occasional shrimp find itself caught up in water hot enough to turn it instantly into bouillabaisse?

It was Peter Rona, a geologist with the NOAA laboratory in Miami, who first discovered hot springs in the Atlantic in 1985 and collected shrimp for biologists to examine. Using a dredge to sample the seafloor 3,600 meters below the surface vessel, Rona picked up hundreds of shrimp and pieces of black sulfide chimneys. Most of the shrimp quickly found their way to the Smithsonian Institution, in Washington D.C., where Austin Williams, one of the world's experts on shrimp, lobsters, and crabs, studied them. Williams and Rona published descriptions of two new species of shrimp, assigning them the names *Rimicaris exoculata* and *Rimicaris chacei*. The generic designation, *Rimicaris,* is from the Latin *rima,* meaning rift or fissure, and refers to the Mid-Atlantic Rift; *caris* means shrimp. The specific name *exoculata* refers to the fact that this species is deprived of any vestige of the usual shrimp eyestalk or cornea; *chacei* is named in honor of Fenner A. Chace, a renowned taxonomist of decapod crustaceans. Both species are members of the same taxonomic family as shrimp that are found in Pacific hydrothermal vents.

The Pacific version of the shrimp live somewhat inconspicuously as ordinary scavengers among groups of other animals at deep, warm water (2–20°C) springs. It is the Atlantic branch of the family that is far and away the more remarkable in terms of its ecology. Spectacular crowds of these shrimp, as many as two thousand per square meter, have been observed completely obscuring the sulfide surface beneath them. They don't just sit quietly, but constantly move about in such a way as to prompt John Edmond, an MIT geochemist who has visited the Atlantic hot springs in *Alvin,* to describe them in his broad Scottish brogue as "disgustingly like maggots swarming on a hunk of rotten meat." While I might have opted for a more engaging analogy—say, bees dancing on a hive—Edmond's imagery does justice to the site.

Another extraordinary feature of the Atlantic shrimp is that they exclusively dominate the fauna at two of the Atlantic hot springs. This contrasts sharply with springs of the eastern Pacific, where lush, exotic communities of tubeworms and bivalves crowd around cracks in the seafloor through which warm water issues. It is the tubeworms and bivalves that have become famous for their symbiotic associations with sulfur-oxidizing bacteria, housed within special tissues and producing most, if not all, of the animals' nutrition. The Atlantic shrimp, however, do not host endosymbiotic bacteria. Instead the shrimp appear to gather their food by mining the sulfide surface of the black smokers on which they live and feeding on filamentous bacteria that foul their mouth parts. The tips of the legs of these shrimp have strong, filelike spines that may be used for rasping. Their first pair of legs, located very near the mouth, have scoop-shaped claws that look well designed for picking up small bits of loosened sulfide; an associated brushlike appendage could sweep the sulfides out of the scoop and into the mouth.

On postmortem dissections of collected specimens, I

found every stomach packed solidly full of sulfide minerals. Of course, there isn't much nutrition to be gained from the sulfide minerals themselves. But associated with the sulfides are tremendous numbers of bacteria. Like the symbiotic bacteria of the eastern Pacific tubeworms, these bacteria grow using the chemical energy in reduced sulfur compounds, which are plentiful in the warm vent water, to convert carbon dioxide and water into bacterial material. Bacteria-laden sulfide minerals are ingested by the shrimp, the bacteria are digested, and the undigestible minerals are eliminated. This mode of feeding would account for the determined way the shrimp seem to attack the sulfide chimneys, as if desperate to glean yet more bacteria from an otherwise unpalatable substrate.

Palatability raises another issue: are the shrimp good to eat? The opportunity to address this issue arose during the visit of a very distinguished and discriminating colleague from the University of Leeds in Great Britain, Professor J. R. Cann. We gathered around the laboratory Bunsen burner one wintry afternoon in Woods Hole, took one of the shrimp from the freezer, and boiled it. It did not turn an appetizing pink. If anything, it turned a still more unappealing shade of gray. As we might have expected, given the sulfitic environment bathing the shrimp, the flesh tasted of rotten egg and, far from being succulent, the texture of the beast was as I imagine a rubber band might be. Perhaps it was overcooked. We concluded that there would be no market for these shrimp, even among the most adventurous of the gourmandizing public.

I studied photographs and videotapes of the shrimp to learn about their behavior and could not help but notice

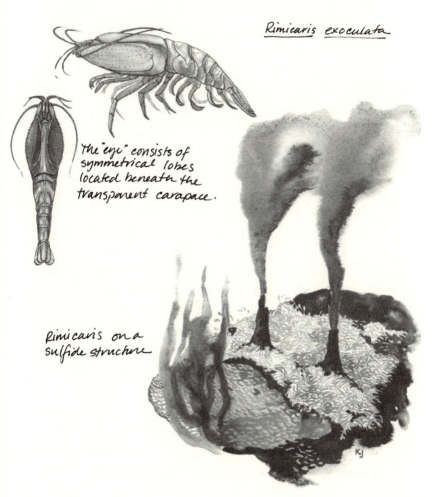

Rimicaris exoculata

The "eye" consists of symmetrical lobes located beneath the transparent carapace.

Rimicaris on a sulfide structure

a pair of bright reflective spots on the dorsal surface, or back, of the shrimp. The spots are not obvious in the preserved specimens I had in my lab, but when I looked carefully I discovered that the spots correspond to the paired lobes of a very large and unusual organ just beneath the thin, transparent carapace or shell of the shrimp. Each lobe was connected to the brain of the shrimp by a large nerve cord. Despite the absence of an image-forming device, I guessed the lobes corresponded to eyes of a sort never encountered before. My guess was hardly proof, as my colleagues were very quick to point out, so I set out to find what was needed to show that they were indeed some sort of eye in this otherwise eyeless shrimp.

The proof required turned out to be the unequivocal demonstration of the presence of a light-sensitive pigment. There are two straightforward ways of doing this: one relies on immunological techniques, that identify molecules on the basis of structure; and the other is a biochemical assay which identifies molecules on the basis of function. Ete Szuts, a sensory physiologist at the Marine Biological Laboratory in Woods Hole, was willing to perform the biochemical assay. Together we dissected the organs from frozen shrimp under the surreal conditions of a red-lit laboratory. Then Szuts purified the membranous material that should contain the visual pigment, and extracted whatever pigment there was with a mild detergent. We used a spectrophotometer to measure the amount of light absorbed by this extracted material, first in the dark and then after bleaching the extract with light. The two measurements are necessary, since visual pigments are light sensitive and have characteristic absorption spectra under these different light conditions.

It is an elegant procedure, producing satisfyingly concrete results when it works, which it did for us. In the extracted material there was a substance that absorbed maximally in the long-wavelength, blue-green part of the spectrum; on bleaching, the product absorbed maximally at shorter wavelengths. The shape of the absorption spectra of the shrimp pigment closely matched those of rhodopsin, the visual pigment found in eyes of both vertebrates and invertebrates.

What are the shrimp looking at? Without lenses, they cannot be seeing an image. Instead we guess that the shrimp are detecting gradients of light. Based on the structure of the organ, we think that it is well adapted for detecting very low levels of light. What sources of dim light are there in the deep sea? These shrimp live 3,600 meters below the surface, far beyond the reach of sunlight. It is a pitch-black environment. From *Alvin,* the only light to be glimpsed at these depths is the occasional, eerily blue-green flash from a bioluminescent animal. Normal shrimp eyes can detect this type of light. Why should the vent shrimp have evolved such an unusual eye if this was all it was looking at? We began to wonder about other sources of light that might be peculiar to the extreme hydrothermal vent environment.

The dominant features of the Mid-Atlantic Ridge vents are the sulfide chimneys on which *Rimicaris exoculata* lives. Could there be light, detectable by the shrimp, associated with the jets of 350°C water? The advantages to the shrimp of such a situation are clear: the light could serve as a beacon to draw them to areas where they can feed, and such a light could also serve as a warning signal to deter them from too close an encounter with water hot enough to cook them instantly.

We know that hot things glow with thermal radiation, a phenomenon known as "black body" radiation. Are black smokers hot enough to emit light visible to the shrimp? Rough calculations, based on estimates of the threshold light intensity necessary for vision, the emission spectrum of a black body radiator, and the absorption spectrum of the visual pigment of the shrimp, indicate that the shrimp may indeed be able to see such a glow, even though it might be too dim for a human eye to detect.

Testing this hypothesis means returning to Mid-Atlantic Ridge vents with *Alvin,* carefully measuring light levels and wavelengths at the chimneys, studying the shrimp's behavior in response to experimental light stimuli, and conducting shipboard physiological experimentation. We did return to study the shrimp's eye, but found it impossible to sample the shrimp without blinding them with submersible lights. Even so, dozens of animals were carefully treated and preserved for a variety of studies that will help us understand how the eyes work.

Logic led us to believe that our hypothetical light at Mid-Atlantic Ridge vents could be a universal phenomenon at similar high-temperature vents elsewhere in the deep sea. Thus, although we could not immediately find out what light *Rimicaris exoculata* might be detecting, we could ask a related question: What are ambient light conditions at black smoker chimneys?

The opportunity to begin answering this question came unexpectedly and quickly. John Delaney invited me to participate as the biologist on an *Alvin* dive series to hydrothermal vent sites on the Endeavour Segment of the Juan de Fuca Ridge, off Vancouver. I had learned that Delaney was to use an electronic digital camera on the impending cruise to create a digital mosaic of seafloor images in the vicinity of the vents. At about the same time, I was reminded that such a camera ought to be sensitive enough to detect the levels of light I expected the shrimp to be seeing. Conventional photographic emulsions would have to have an ASA rating on the order of 50,000 to 100,000 to detect the same level of light. High-tech digital cameras are used extensively in astronomy to capture dim light from distant galaxies. There is a satisfying, if somewhat pre-Copernican, symmetry in turning to the same technology in oceanography to capture light emissions fueled by the core of our own planet.

On reaching the *Atlantis II* and finding Delaney, I told him about my shrimp eyes and what I thought about light at vents. I suggested aiming the camera at a black smoker orifice, extinguishing *Alvin*'s outside lights, and letting the camera record the ambient light. Delaney was enthusiastic, and Dudley Foster, *Alvin*'s chief pilot and expedition leader on that cruise, agreed to attempt the experiment.

What initially sounded like a simple experiment in fact required a great deal of effort. Together with Milt Smith, an expert in remote sensing from the University of Washington, the *Alvin* crew worked into overtime to configure the camera so that it could collect the data.

Finally, on the last dive of a nineteen-dive series, *Alvin* was lifted off the deck carrying the CCD camera mounted on the front of the basket. Inside the sphere were Foster, Delaney, and Smith. That day I haunted the top lab where communications with the sub take place every half hour. In response to the brief surface queries about their status, the sub replied with a "busy" signal in Morse code. At the end of the dive, with the submersible well into its hour-long ascent, I gave up on learning anything about the success of the experiment and left the room. On returning, I was handed a note by Pat Hickey, the dive's surface controller. It was a message finally relayed from the submersible, a message with only two words: VENTS GLOW.

With *Alvin* on deck, scientists and pilot gathered around the computer workstation as Smith recalled images of the glow. I expected to see some ambiguous hint of a fuzz, which, if one was willing to stretch the imagina-

tion, might be called a glow. I doubt that I was alone in
that expectation. But what came up on the screen instead
was a dramatic, unequivocal glow with a sharply defined
edge at the interface between the sulfide chimney and the
jet of hot water. Just 1 or 2 centimeters above this inter-

face, the glow became very diffuse, disappearing altogether within 10 centimeters. The same phenomenon was documented at two different chimneys within the same vent field.

The discovery of this glow at high-temperature vents opens up a whole new area of research. At the moment, the glow is an intriguing and aesthetically pleasing phenomenon. Its importance will be judged by what we will learn in the future about the mechanisms of its production and its biological consequences.

10

BIOLOGICAL
PARTICULARS

T HE ATLANTIS II *IS NOT A LARGE SHIP. AT 210* feet, she is easily dwarfed at the dock by other vessels. But she has proud and distinctive lines, with her classic bow, maze of superstructure, and distinctive stern A-frame. For the ship's crew and the *Alvin* Group, she is home. For some, she is the only home. There are a few sailors—the old-timers, mostly—who seem to have given up on life onshore. It is easy to understand why. Life on the ship involves no rent checks or utility bills. There is little need for decent clothes or a car. Serious problems from a life onshore can fade rapidly to distant memories as sight of land is lost. For some sailors, life on the *AII* is an escape.

R/V Atlantis II

The *AII* is an old boat, built in the early 1960s, and she is just a year from her retirement when she will be sold to the highest bidder, or, more likely, nearly given away for scrap. *Alvin* will transfer to yet another mother ship. I see a nascent nostalgia aboard the *AII* during the last year of her life as her sailors prepare to move to another ship. Because I sail so often and was part of the crew myself for a couple of years, the crew of the *AII* is like family. Over the past decade, I have watched the progress of careers as mess attendants become ordinary seamen, cooks become stewards, oilers become wipers, engineers become chiefs, able-bodied seamen become mates and mates become masters. I watched as the ship took on women, first in the galley but soon joining the crew to work the decks, the engine room, the bridge.

On long cruises, it is the steward's department that can make or break a leg. Meals these days are excellent and require a daily workout in the exercise area to keep from acquiring unwanted pounds. It is rare to pass by the stepper, bikes, bench, and set of free weights that make up the exercise gear and not find one or another of the crew working out. The mess deck doubles as the movie lounge, with a video library housing thousands of films. There is a well-stocked book library too. The library is a quiet spot where it is all too easy to be lulled to sleep by the roll of the ship. I prefer to spend odd moments on the bridge. It is always quiet and orderly there, with a maritime standard of etiquette. At night, when the bridge is dark and manned only by the mate on watch and a helmsman, it is a good place for thought and daydreams, both of which I need if I am to work through weeks at a time at sea.

In her design, the *AII* is unmistakably a work boat. Farther forward on the main deck from the A-frame is the *Alvin* hangar, where the submersible is serviced every evening. Adjacent to the hangar are the *Alvin* shops—two small cubicles loaded with supplies and spares, instruments and tools, for maintaining the sub. The main science lab is a flexible open space that looks more like a small flea market arena with its plywood benches than it does a scientific theater for cutting-edge oceanographic research. But once a science party comes aboard, computers and electronics and chemical analyzers and charts occupy all of the usable space on the benches and spill over onto the deck. Space wars, where investigators bicker over a couple of extra square feet of bench, are a customary ritual prior to a cruise. Often entire university laboratories, shipped in dozens of crates to remote ports, are assembled and disassembled before and after each science leg.

A busy main lab crowded with scientists is emptied in short order when the bridge passes an alert over the squawk boxes that are found throughout public spaces on the ship: "Whales off the starboard quarter!" We all drop what we are doing and scurry to the fantail. There, halfway to the horizon, we watch: a spout of water is exhaled; a dark body breaks the surface of the sea and then slips like a comma beneath; then nothing. I used to grab my camera whenever I heard a whale alert, in anticipation of capturing a shot of a whale breaching, water droplets sparkling in the sunshine as the whale collapses back onto the sea. But my pictures of wild whales sighted from the ship invariably turn out to be shots of sea and sky, the whales invisible. Now I leave my camera behind and am more philosophical: if by chance a whale happens close by, I will watch the whale pass and hold the moment quietly within.

Although I have never seen a live whale from *Alvin*, I have seen a dead one on the seafloor. The flesh is long

since gone from this whale, having undoubtedly provided
a brief but sumptuous repast for opportunistic scavengers
of the deep sea. But its white skeleton, all 20 meters of it,
lies neatly on the surface of a bed of cold, tan sediment,
in perfect anatomical order—tailbone to massive jaws,
vertebrae stretched in linear order between. It is the size of
the mandibles—more than twice as long as I am tall—and
their distinctive, curved shape that indicate in a glimpse
the abyssal skeleton belongs to a baleen whale, probably a
blue or fin whale.

This whale skeleton was discovered several years ago
by a group of scientists working a deep-sea station in
1,200 meters of water, off the coast of California. Paul
Tibbetts was the pilot. Always a kind of rogue pilot, Tibbetts told me he was bored of driving the photo transects
prescribed by the onboard scientists and veered just a little

whale lies on soft mud
which is easily disturbed.

Vertebrae are covered
with bacterial "fur" and
small invertebrates

one meter

CATALINA BASIN
WHALE SKELETON

bit off course to take a look at a sonar target that he thought looked maybe like an old wreck. As he approached the target, Tibbetts was convinced he'd found a dinosaur skeleton, but a closer look proved it to be only a bit more prosaic.

What happens to a large fish or a marine mammal when it dies? It will most likely sink to the seafloor and provide an intense and localized enrichment of food for the normally famished benthic community. Large scavengers quickly move in to attack such a food-fall, tracking the scent of dead flesh up the downstream plume. Bottom-feeding sharks show up to gnash at the meat. Giant amphipods, the size of rats, appear out of nowhere to nibble on the remains. Brittle stars form tangled mats of skinny arms as they gnaw their way through the flesh with miniature bites. Soon, all that is left is a pile of bones.

Bones of whales are permeated with oil. Museum curators know this because whale bones stashed on archival shelves will weep lipids for decades. The Catalina skeleton on the seafloor is slowly releasing its lipids. When I dove on the Catalina skeleton, I saw white filamentous bacterial mats growing on the exposed surfaces of the bones and over the nearby sediment. I brought back one of the vertebra for Craig Smith, an oceanographer from the University of Hawaii who is studying the skeleton. The vertebra smelled of sulfide and of a peculiar rankness that I could not identify. Small mussels, limpets, and serpulid polychaetes were picked off the bone with fine forceps. I also sampled the nearby muds where white-shelled clams live and where brittle stars browse. Craig finds that many of the species living on the Catalina whale are closely related to vent species and that they derive their nutrition at least in part from chemosynthetic bacteria. Those lipids oozing out of the bone are being decomposed by bacteria that produce hydrogen sulfide as a by-product—sulfide, the compound that fuels chemosynthetic bacteria at hydrothermal vents. No one knows for sure how long the bones have been lying there, nor how long they will continue to support the sulfide community.

It always amazes me that animals reach the surface alive from great depths. The 4-inch thickness required of *Alvin*'s titanium hull and acrylic viewports attest to the tremendous pressure exerted by the overlying water at depth; as do Styrofoam cups, artfully decorated and cycled to and from the seafloor tied to the external frame of the submersible, a cycle that permanently shrinks coffee cups to shot-glass size. Shrunken Styrofoam heads are a special favorite, the heads taking on alien features as air squeezes out of the foam. Mine is a shrunken blue bust that occupies a place of honor on my bookshelves, his catlike green eyes with orange slits locked in a perpetual glare. I keep it within easy reach to show inquisitive guests what 3,600 meters of water can do to an air-filled bit of foam.

Surprisingly, although decompression must place tremendous stress on an animal's physiology, it seems that temperature shock is the proximal killer. On the seafloor,

I have carefully collected specimens so they are alive and unharmed and placed them in an insulated container flooded with cold bottom water. Close the lid before you begin your ascent and the water will stay cold—near 2°C. Usually at least some of the animals will survive the transit, despite a pressure change of several hundred atmospheres. These same animals exposed to ambient temperatures during the ascent, which can often change by as much as 30°C over a two-hour period, will be dead when they reach the deck. Even animals that normally live in warm hydrothermal waters survive only when kept chilled.

It is only animals that lack gas bladders or other gas-filled organs that survive decompression. Gas in the swim bladder of a fish expands as pressure decreases on the ascent until the bladder bursts and the fish dies. But tube-worms, polychaetes, crustaceans, and most other invertebrates have only fluid-filled spaces between tissues. Tubeworms are pretty fragile, surviving on deck for a few hours if kept chilled, a few days if recompressed in special high-pressure aquaria. It is difficult to re-create in the laboratory the chemical environment in which they thrive. Tubeworms need flickering sources of sulfide and oxygen, a condition that is difficult to achieve artificially. When they first arrive on deck, if left unjostled in baths of refrigerated seawater the worms will slowly poke their plumes out of their tubes. Brush a plume lightly and the animal instantly retreats into its tube.

Vent crabs are among the toughest of the lot. They almost always come up alive, although decompression affects their neuromuscular control so that they stagger about like drunken sailors rather than scuttling sideways as they normally do. Put a crab back under pressure and it will live indefinitely so long as you drop in a morsel of meat every so often.

One of the most common vent crab species is, as an adult, about the size of a rock crab like you might find camouflaged among brown fucus weeds on a rocky New England shore. In the deep sea, where there is virtually no light, camouflage is not necessary. The carapace of the vent crab is porcelain white suffused with lavender, the claws manicured black at the tips. The white crabs are easy to see against the black of lava and, since they range beyond the border of the vent field, they are one of the first signs of an approaching vent in areas where vents are common. The crabs just as often wander into warm water among tubeworms, seeking out smaller animals on which to feed—limpets and polychaete worms seem to be favorites, judging from the few stomach contents I have examined. When they can catch them, the crabs will nip off bits of red plume from tubeworms. And when *Alvin*, not always the most delicate of collecting tools, damages tubeworms, the crabs quickly sense the wounds and pile up on the hapless worms to feast on what must be, to them, succulent flesh.

It is the megalopal stage of a vent crab that breaks

records for tolerance to decompression. Crabs of any kind, shallow-water or deep, pass through a series of life-history stages. A female vent crab produces prodigious numbers of small eggs that she broods on pleopods underneath her abdominal apron. When the eggs hatch, tiny, swimming zoeal larvae are released into the water. The larvae are so different in form from the adult crab that you would not guess from one that you could derive the other. They are transparent, with a bulbous anterior end armed with forward and lateral spines and a long, skinny abdomen divid-

Megalopa

Intermediate developmental stages of vent crab

Actual size: 1 cm

Bright orange and yellow pigmentation

Abdomen segmented w/ "swimmerettes" used for locomotion

Megalope clasps claws and legs to body to scoot through water rapidly

KJ

ed into segments and ending in a forked tail, all quite small and delicate. The zoea swims weakly using tufted pleopods arranged in pairs on each segment of the abdomen. Like the larva of a mosquito, the zoea of a crab passes through a series of instars in the water column as it drifts with the currents. The larva grows incrementally, shedding its tiny exoskeleton like a discarded dress and then swelling quickly to a new size before the fresh exoskeleton has a chance to harden. After a number of these molts and bursts of growth, the zoea undergoes a metamorphosis, transforming into a megalopa.

The megalopa is recognizably crablike, except that it can both swim and walk. It is the megalopal stage that makes the final choice of where on the seafloor it will begin its benthic life since, with the next molt, the megalopa becomes a juvenile crab and loses its ability to swim. A peculiarity of the megalopa of the vent crab is its size. It is large—giant—at least for a megalopa, which is to say that the megalopa of a vent crab is the size of my fingernail. And it is bright orange. These two characteristics combined make swimming megalopae easy to spot from the submersible. Where the adult crabs are particularly abundant, as on portions of the East Pacific Rise, I have counted dozens of megalopae in the water column, sometimes tens of meters above the seafloor and hundreds of meters from any known vent.

It happens that tubeworm thickets are refuge for megalopae so that, in collecting a clump of tubeworms, we get megalopae as a bonus. Megalopae are tough. I have kept them alive in glass beakers, stowed in a refrigerator—at sea-level pressure, 1 atmosphere—for weeks. Although eventually the megalopae stop swimming and show signs of physiological stress, they are active and well coordinated for the first week or so . Their hardiness, their barotolerance, must derive from the flexibility they need to locate a suitable site to metamorphose. Since hydrothermal vents are like islands, discrete and disjunct, the megalopae must be good at moving up and down in the water column, searching for a chemical plume to guide them to a vent site. To move vertically, they have to tolerate pressure changes. Once they lock into a benthic existence and become juvenile crabs, they no longer need to be barotolerant; their physiology becomes modified to accommodate only a very narrow range of pressure.

There is a bittersweetness to sampling animals from the deep sea with *Alvin*. At once there is great delight in knowing that perfect specimens will be available for study—that the animals are often alive and responsive when they reach the surface—and certainty that the animals will soon die. I have often wished for biosensors like those that Dr. McCoy carries on the Starship *Enterprise:* noninvasive instruments that reveal the physiology and ecology of unusual life forms without sacrificing the organism. But instead we rely on collection, dissection, and assay to tease out the biological particulars of an animal.

11

THE TRILOBITE
FACTOR

I LIKE MAPS. THOSE OF US WHO ARE TRAVELERS
and adventurers on this planet pore over maps as if
they are Rosetta stones, with the key to knowledge and
understanding lying . . . somewhere.

Walk the corridors of an oceanographic institution
and you will find the walls plastered with maps—maps of
coastlines and continental shelf bathymetry, sea surface
temperatures, ocean surface currents, and deepwater circulation patterns; maps of chemical anomalies, animal
distributions, sediment type. They come in all varieties:
three-dimensional, with those funny glasses pinned to the
board by a string; two-dimensional maps perspective
with vibrant colors—abstract and worthy of a SoHo gal-
lery; dusky side-scan sonar maps—acoustic chiaroscuro;

seismic profiles of the ocean crust—Rorschach arrays of squiggly lines best viewed from a distance with a squint.

You can tell immediately the particular kind of scientist in an office by the maps outside the door. Down the hall from me, Room 224 houses a geophysicist, the entryway framed with maps of mantle Bouger anomalies; outside Room 218 a structural geologist plots locations of volcanic cones she has discovered along the Mid-Atlantic Ridge. In Room 222 resides a hydrothermal vent scientist whose detailed map of hydrothermal sites on a small stretch of the East Pacific Rise reveals his trade, while a child's drawing taped on the door reveals his pride.

I am never far from a map of the seafloor. Shades of blue march in tight bands down the continental slope, light aquamarine deepening to turquoise over the broad Hatteras abyssal plain. Oceanographer, Hudson, and Hatteras Canyons crosscut the slope, draining shelf waters from their headlands off Cape Cod, New York City, and Cape Hatteras into the deep Atlantic. The canyon walls are steep and irregular and, although I have not dived these canyons myself, I know them to be prime coral habitat. I have a severe bias in my coral appreciation standards. The showy, luxuriant corals of the great tropical reefs are certainly precious, but to me they seem overdone, the floozies of the marine invertebrate world, rouged and primped beyond my interest. I prefer the more austere skeletons of the corals that live frugal, ancient lives in the deep sea. In the canyons, coral fans stretch into the current from rocky outcrops, branches lightly encumbered with slender waving fingers of brittle stars.

A chain of ancient seamounts stretches east from Georges Bank, off Cape Cod. On my map they look like submerged stepping-stones for a giant King Neptune, each stride 100 kilometers. Steep-sided and rugged, the seamounts would be a difficult climb were they on land. To the south, the stately Bermudas rise. It is to Bermuda that *Alvin* goes for her certification dives following her periodic refits. The sub needs to dive safely to her full depth capability after a major overhaul in order to win her navy approval; Saint George offers easy and close access to the requisite deep waters. The deepest waters in the North Atlantic, where the color on my map deepens to cobalt, is the crescent that hugs the north shores of Puerto Rico and the Leeward Islands—the Puerto Rico Trench. Seafloor plunges into this 7,000-meter trench—subducted, gone.

But these features are just small details on my map. In your first glance, your eye must be drawn immediately to the band of high, light blue relief that snakes through the exact center of the Atlantic Ocean. It is a mountain range that dwarfs any subaerial system. In the jigsaw puzzle of the earth's tectonic plates, the western flanks of this mountain frontier stretch toward North America as one piece of the puzzle; the eastern flanks that head toward Europe and Asia are another piece. Follow the Mid-Atlantic Ridge north. It takes a major jog to the west at the Charlie Gibbs Fracture Zone, a deep transverse cut at the latitude of Ireland's County Cork. Then the ridge makes a shoaling beeline to the east-northeast, straight up to Iceland along the Reykjanes Ridge. Here the mid-ocean ridge has gone terrestrial.

If you go to Iceland and find your way past the Icelandic ponies to the ancient village of Thingvillir, and then if you stand astride a particular small fissure, you will be astride the North American and Eurasian plates. Shut your eyes in this place and you might almost come to terms with the notion of an ocean basin that continuously opens and closes over that unfathomable geologic time. If you wish a sense of scale and landscape of seafloor proportions, wander about the treeless biome of Iceland and seek out faults, horsts, blocks, and fissures, and search for wild hot springs.

I have traveled into the waters north of Iceland, searching for vents on the remote Mohn's Ridge, a northern extension of the Mid-Atlantic Ridge, well above the Arctic Circle. I went there because the Arctic Ocean is the youngest basin in the Atlantic. Its deep waters, and thus presumably its fauna, have been hydrographically isolated from the main Atlantic basin by twin, shallow, east–west sills—the Greenland-Iceland Rise and the Iceland-Faero Rise to either side of Iceland itself—for as long as the Arctic has existed. I wanted to know what kinds of animals have managed to colonize sequestered, putative hydrothermal systems there.

In summer weeks of calm seas, long arctic days and brief nights, we plied the ridge with a towed sled armed with cameras and lights looking for hot springs. It was as if we were searching a wilderness from a helicopter for a particular bit of uncommon ground—at night, with no sight except for a video image telemetered from a camera dangling beneath us on a mile-long tether, the ground illuminated by a powerful flashlight. Although we learned much about the geology of the seafloor, the hot springs eluded us. I thought for a while that white deep-sea sponges might be marking the way—there were areas where the sponges were concentrated, maybe a sign of a nearby vent. But we found nothing, save some soft yellow staining of sediments.

Late in the cruise, we shifted position farther to the north and east, targeting a shallow seamount where an eruption was reported to be ongoing. Warily we skirted pack ice, probing for an alley that would guarantee us safe passage to our target. But our ship was thin hulled, unfit for duty as an icebreaker, and, finding no clear avenue, we were forced to abandon the effort. Disappointed and out of time, we steered the course for Reykjavik. Our mood lightened as we passed the island of Jan Mayen, which obligingly lifted its kilt of clouds to offer a rare and stunning glimpse of its volcanic silhouette. With our failure to find the vents, I am left to wonder still if the arctic seas are refuge to novel forms of vent life.

The Mid-Atlantic Ridge bisects the oldest part of

the Atlantic Ocean, above the equator. Where the magma that forms the ridge buds off laterally, the oceanic islands of the Azores are found. A younger part of the Atlantic Basin results from the separation of South America and Africa, which began at the southern Capes, causing a single rift system to propagate north through the southern ocean. North and south ridges met at the Romanche Gap, just at the equator, and are now contiguous as one long spreading center that runs uninterrupted by land all the way from Iceland south toward Antarctica, where it sweeps broadly to the east and stretches into the Indian Ocean.

While it is hardly fair to claim that the northern part of the Mid-Atlantic Ridge has been "explored," since we have seen so little of it, the southern Mid-Atlantic Ridge is truely virgin. *Alvin* has never dived in the southern Atlantic, nor, to my knowledge, has any other deep-diving submersible. *Alvin*—this is so unbelievable that I denied it the first time I heard the claim—has never dived south of the equator, not in any ocean. I have been with her as she worked beneath the Southern Cross very near the equator, close by the Galapagos Islands, diving on Rose Garden. In these waters, her mother ship will sometimes take the sub and crew on a twelve-hour run due south. The ship turns about to north just as her bow breaks the line that marks the equator, giving all Pollywogs on board the chance to become Shellbacks in "secret" maritime rites of initiation—humbling homages to King Neptune—that

mark a sailor's first crossing of the equator. I am a Shell-back of long standing now. I carry my Shellback card with my passport as proof, should I happen upon a ship that threatens me with reinitiation.

It is to the remote oceans, the Arctic, the Indian, the southern oceans, that I look for vindication of what I think of as the Trilobite Factor—the belief that there are still major discoveries waiting for us in unexplored regions of the ocean. Trilobites, of course, have been extinct since the Permian; chances of finding some deepwater relative of these fossil arthropods teeming about a deep-sea hot spring refugium are perhaps remote to nil. But, if not trilobites, surely other, unimagined celebrations of na-ture's ingenuity will be found. Mounting an expedition to a remote field location requires years to coordinate, but those of us who take on such time-consuming and thank-less administrative tasks know that the rewards of explo-ration and discovery make it all worthwhile.

12

SEDIMENT PLAINS

I WAS NOT AT ALL PLEASED AT HAVING TO JOIN
the line outside the ship's hospital, waiting my turn for
a gamma globulin shot from the medic. But I was on
the schedule as swimmer for the launch the next day,
which meant that I had to ride on top of *Alvin* as it was
lifted off the fantail and lowered into the water, in order to
help handle the lines. Swimming is usually a welcomed
chore, an opportunity to abandon the ship for a few minutes and get into the water. This time, though, less than a
quarter of a mile away from our ship, a barge carrying
sewage sludge from the New York–New Jersey metropol-
itan area was flushing the collective toilet. The sea around
us was brown and smelled as if we were within a mile or

two of shore, when in fact we were 106 miles due east of the casinos of Atlantic City.

The gamma globulin boost was preventive medicine against the filth of the sludge. With the barge as our temporary companion, we floated within invisible boundaries drawn by government agencies that define Deep-Water Dumpsite-106. DWD-106 lies just at the deep edge of the continental slope, where water depths of 2,000 meters gave planners a sense of security. The sewage sludge, they reasoned, would be diluted and dispersed in surface waters and would never sink through the hundreds of meters of water column to be measured in any detectable amounts on the seafloor. And so, with flawless bureaucratic logic, monitoring efforts focused on samples from the upper water layers.

I think nearly every deep-sea biologist has a dog-eared copy of the New Yorker cartoon by Charles Saxon which appears on the following page. As a pilot, I cannot wax too poetically about diving on a soft-mud bottom, which I must suspect this woman had in mind as she expressed her indifference about the deep sea.

Mud bottoms have an obvious disadvantage from a pilot's point of view: the sediment surface comprises very fine, very loose particles, and is easily disturbed. One false move with the submersible and you find yourself lost in a cloud of sediment. In deep waters, where bottom currents are often sluggish, the cloud takes forever to move away. If one patch of mud is as good as another, you can outrace

*"I don't know why I don't care about the bottom
of the ocean, but I don't."*

Illustration by Saxon; © 1983 The New Yorker Magazine, Inc.

the mud cloud to settle the sub in a clean patch upstream.
But if science dictates that you must work a specific spot,
you can do nothing but keep the sub absolutely still and
wait patiently for the cloud to drift away. Many peanut
butter and jelly sandwiches have been eaten a mile and a
half below the surface while waiting for the water to clear.
Where we do work, we leave behind a distinctive scar in
the mud traced by the front ski of the basket as it skims
over or rests on the sediment surface. The trace lasts for

years and bears witness to the slow pace of change in much of the deep sea.

The generation of deep-sea biologists just before mine cut their teeth on the organisms that live in soft-sediment environments. There were only a handful of these scientists in the United States, but their work and ideas continue to influence the field. At one time, Woods Hole was host to the three most prominent deep-sea statesmen—Howard Sanders, Robert Hessler, and Fred Grassle—who were the first to appreciate the extraordinary richness of species that live in the soft sediments of the deep sea. They are gone from Woods Hole now: Howard to retirement; Bob to Scripps in California many years ago, where, in partial protest of scarce funding for deep-sea biology, he now studies details of the biology of a primitive shallow-water crustacean; and Fred, whose work extolled the rich diversity of the deep sea, to a prestigious director's chair at Rutgers. I can count on one hand the number of biologists in the United States who are currently totally committed to understanding the biology of the deep sea the way Sanders, Hessler and Grassle were.

The concept of a rich deep-sea biodiversity has been hard won through the very unglamorous processing of mud collected from the seafloor. Just developing a box core that an *Alvin* pilot can easily use to obtain a reliable, quantitative sample took some careful engineering. The concept is straightforward: the pilot picks up a stainless-steel box core from the sample basket, gently pushes it into the mud without disturbing the surface, and then fires spring-loaded doors that close down on the bottom of the core, securing the sample. The trick is to design a door system that will not misfire during deployment yet will fire reliably when activated. More than once I have carefully positioned a core and fired the springs, only to have the doors remain open. Then you have to use the fingers of the manipulator to shut the doors manually. A pilot's transcript at this point becomes rather blue. I confess even my ordinarily gentle vocabulary could give way to some pretty fluent swearing.

Once a core is on deck, it is processed by methodical sieving through fine-mesh screens, the objective being to eliminate as much mud as possible while retaining all of the organisms. I am always impressed at how cold the mud is and at how old that cold feels. The organisms are generally so small as to be unnoticeable. Although the diversity might be rich, productivity of most of the seafloor is low and biomass of organisms picked from the sediment samples is moderate. Samples of mud and organisms are preserved in jars to be looked at and sorted and counted and identified on shore under microscopes.

I don't know the exact surface area of the deep seafloor that has been sampled in this manner and examined for organisms, but I think it is a safe bet that it would barely begin to cover a football field or two. The audacity of our extrapolation of any concepts based on these samples to the seafloor as a whole is pretty supreme; the meagerness of samples argues only that there are volumes we don't begin to know or understand about deep-sea ecology. Bear in mind that I am not talking about ignorance of an esoteric ecosystem; the seafloor and the water column above it comprise the largest ecosystem on our planet.

We think of the deep sea as a food-limited environment. Absorption by seawater extinguishes visible light below a few hundred meters, so deeper waters cannot support primary production by photosynthetic phytoplankton. This means that the food supply to the seafloor must all be generated in the upper, lighted layers of the water column. To reach the seafloor, a particle of food runs a gauntlet of consumers that take their share of the organic material from the particle. Through each cycle of consumption, the nutritive value is reduced. Particles that sink slowly may be recycled many times, so that what finally reaches the bottom is of pretty poor quality. Sometimes conditions may be right for particles to sink rapidly and relatively uncycled: maybe the die-off from a surface bloom of phytoplankton overwhelms consumers and falls as a pulse of food to the bottom; or maybe a raft of gelatinous zooplankton near the sea surface succumbs to a change in temperature or a seasonal paucity of food and slips rapidly in death through the water column.

Or maybe sewage sludge dumped on the surface sinks faster than anticipated and becomes available to the bottom dwellers. After the dumping at DWD-106 was approved, computer modeling of newly available data suggested that there could be a fallout shadow of sludge particulates accumulating on the seafloor downstream of the dumpsite. A group of academic scientists, led by Fred Grassle and relying primarily on *Alvin* for access to the seafloor off the coast of New Jersey, showed this indeed to

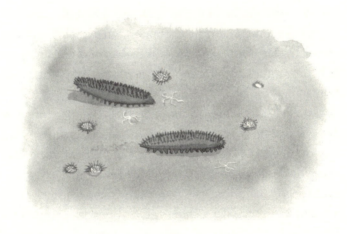

be the case, and that the organic material was entering the benthic food web. The effects are subtle. Dive in *Alvin* to where benthic accumulations are expected to be greatest, and there is perhaps a sense that more urchins and cucumbers graze the surface muds here than in a control area upstream of the dump site, at the same depth. I have dived in the downstream shadow and could not detect any noticeable layer of sludge on the seafloor. Bottom muds do not come up smelling of sewage. The only clues to contamination come from laboratory analyses of sludge-associated chemical signatures like silver and polyaromatic hydrocarbons and linear alkyl benzenes, by-products of human civilizations.

What is to me an inconceivable volume of wet sewage sludge, more than 6 million metric tons per year, was dumped at DWD-106 during the years of legal offshore disposal, making this arguably the largest explicitly sanctioned pollution of our blue-water oceans. The amount of material reaching the seafloor probably doubled the normal flux of organic material derived from photosynthetic primary production by phytoplankton of the overlying surface waters. This almost certainly has had an effect on the reproduction and growth rates of the animals dwelling on the seafloor, an effect that is just now being assessed by colleagues in the United Kingdom. Yet we were naive enough at the start to dismiss the idea of monitoring the seafloor altogether during this experiment in oceanic-scale eutrophication.

The oceans have a large but limited capacity for abuse. It doesn't bother me so much that the deep-sea benthos received an extra dose of food, although perhaps as a consequence the community carries a burden of assimilated heavy metals and synthetic compounds that could compromise the success of some species' populations. In a contrary way, I like to think that perhaps I personally contributed to the reproductive success of a deep-sea urchin by that flush of a toilet when I was visiting New York City some years ago now. What does bother me is that the seafloor was not monitored. I don't think this was so much a result of any bureaucratic failing as simply a measure of our ignorance about how the deep-sea ecosystem works. Maybe it is too much to expect us to care about the healthy functioning of a seafloor wilderness that is out of sight and populated for the most part by small organisms that don't even all have names and that we can only think at the moment are useless.

Dumping at DWD-106 and everywhere in the open ocean is now banned, not from any awakening of sensibilities but from the more pragmatic disgust and economic poison of unrelated trash that washes up on our beaches. I am oddly relieved to know that my out-of-sight urchins wander the sediment plains once more in the accustomed famine of the deep sea.

Epilogue

I WOULD LIKE TO WRITE THAT THE FLOOR beneath the open sea is beautifully pristine, a place where one is for once removed from the impact of humanity on nature, a place clean of the litter and debris of modern life. But even in the deep sea we leave our spoor. The worst I have encountered was in a small hollow at just over 1,900 meters depth in Guaymas Basin, halfway up the Gulf of California—Steinbeck's memorable Sea of Cortez. White plastic bags of trash had settled into the leeside of a stand of slender, branched black coral nippled with minute fleshy polyps. The bags were frayed and their contents spilled onto the mud. I remember feeling shame at the desecration of something so beautiful. I suppose the bags were dropped over the side of some ship passing by. These days it is illegal to throw plastic over the side, but all manner of other trash is dumped by passing ships.

In popular spots for scientific research, the burden of trash is building. For her part, the *AII* conscientiously sails several kilometers from her dive site on her evening trash runs, so the insult to the seafloor is usually out of sight. The *AII* complies with the law and keeps all of her plastic on board until she returns to port. I quickly become so used to separating plastics from other trash on the ship that when I am back onshore I hesitate and feel guilty as I throw nonrecyclable plastic into a regular trash can. Despite the careful disposal of trash, it is not uncommon to drive past a beer can on any dive. Pilots like to pass close by to read the label and to guess at its source—does it speak of a party on an American ship? Or maybe Japanese, French, or Russian? Even at a new site where no submersible had ever been before, we manage to leave our mark: an *Alvin* pilot was startled to fly over a toilet at Lucky Strike on the Mid-Atlantic Ridge. It had been tossed over the side the night before by the deck engineer.

At other sites on the seafloor, where science experiments are under way, the gathering piles of anchors and cables and abandoned instruments and dive weights are clear signs that we have been at work. I like to think that the next generation of explorers will look on our rusted and corroded technological refuse as the great historical leavings of early pioneers rather than the careless insult to a wilderness that the refuse might arguably be labeled.

Fortunately, most of our leavings will rust away. There is also a growing awareness among scientists that as the effort to study the seafloor increases, so does the need for responsible stewardship.

We are only beginning our attack on the blue waters, with the coastal margins and continental shelf doubling as our beachheads and communal sewers. I think we could screw it up badly—maybe not in this generation, but give us time. We will turn vents into tourist attractions until we kill off all the tubeworms. Then tourists will have to pay to dive to where exotic life forms used to live. In their place they will see plastic purple placards with a raised impression of a tubeworm. Simple text written at a third-grade level will tell of what wonders used to be present and reminisce about the past when tubeworm populations were so incredibly dense they were described by their discoverers as lush oases of life.

Maybe the deep sea, for all its vastness, has no value in our environmental economy. Maybe we could kill off every organism down there and suffer no consequences, lose no biological novelty of human interest. But we can't be sure. We know only the rudiments of deep-sea science. If we don't at least keep track of what we do in deep water and have the option of curtailing or modifying our actions in response to defined measures of unacceptable effects on an ecosystem, how will we know when we have reached the ocean's carrying capacity? How will we know when to back off? It is an easy thing to off-load our wastes at the surface, but it would be quite another feat to attempt remediation of the seafloor if we let things go too far.

In the sediments of the great ocean plains, the scale of life can be minute, generally measured in millimeter- and centimeter-dimensioned, obscure animals that delight the scientist but do little to titillate the general public. Not much is known of this fauna beyond lists of names and tallies and graphs of numbers. Yet the healthy functioning of these benthic communities, perhaps trivial to you and me, must be critical to the balance in the world's oceans when multiplied by the unimaginably vast surface of the seafloor they occupy. Like rain forests, the deep seafloor ecosystems are likely to be fragile, where even small perturbations in the normal functioning of the systems may result in profound consequences.

What still seems remarkable to me is that less than twenty years ago we could not even imagine the existence of hydrothermal vent communities, and we knew next to nothing about the biota that wander the abyssal plains. We landed a man on the moon nearly a decade before we ever saw the heat of our own ocean's crust exhaled through black smokers. Our forays into the deep sea may seem lilliputian when compared with the space exploration program, and by NASA standards the resources of oceanographers are meager. But there has never been a more exciting time to be a deep-sea scientist, and I for one celebrate the fact that the abyssal wilderness is modest neither in measure nor in mystery.

Index

Abyss 12
Acoustic tracking 33, 38, 90
A-frame operator 24
Alice Springs 60
Aluminaut 30
Alvin 5, 16, 18, 21–22, 30
    acoustic tracking 33, 38, 90
    acrylic viewports, melting point 107
    ascent 37, 136
    Atlantis II 5, 90, 135, 139–40, 170
    ballast 33, 37–38, 100, 108, 170
    certification dives 153
    communication 15, 37

depth of dives 15, 29, 32, 38, 94–95
descent 37–38
Galapagos Spreading Center 57, 75
gathering tubeworms 77–78, 146
going dead boat 50
Group 18, 27–28, 42–43
and H-bomb 30
interior seating arrangements 33–35
Lulu 15–16, 30
maneuvering at black smokers 103–05
manipulators 31, 78–79
mud collection 162–63

Operator's Manual
36
passengers 35–36, 42
pilot-in-training
(PIT), 23–28
pressure specifications
34
repairs and overhaul
22–23, 30, 153
safety and backup
equipment 36–41
sample basket 54
and sewage sludge
165–66
sinking during launch
30
and *Thresher* 30
and *Titanic* 30
titanium sphere 37
transponder 89–91
underwater telephone
(UQC), 36–37, 40
water temperature
probes 105
wireless acoustic link
15
Alvinae 60
Alvinella 60, 64
Alvinella pompejana 64
Alvinellids 63, 122
Alvini 60
Alviniconcha 60
Alvinocaris 60
Alvinus 60

Ambient undersea light 32,
133–38, 164
sunlight, absence of
56, 133
Amphipods 67–69, 144
Anemones 55, 96–98
Angel Rock 108
Anhydrite chimney
121–22
Arctic Ocean 154
ARGO-2 58
Ascent, *Alvin* 37, 136
Astoria, Oregon 28
Atlantic Ocean 28, 152–53,
156
Atlantis II 5, 90–91, 139
amenities 141
crew 140–41
meals 141
trash runs 170
Axial Seamount 60, 119
Azoic zone 11–12

Bacteria 56, 81–83
Floc Site 118
and shrimp 129–30
at volcanic eruption
116, 123
on whale bones 144
Balanus 10
Ballast 33, 37–38, 100, 108,
170
Barbecue 116

Basalt 55, 57
    dandelions 69–70
    mussels 84
    pillows 94–95
    spaghetti worms
       69–71
    tubeworms 77
Bathtub rings 50
Beebe, William 7
Bermuda 4, 153
Black body radiation 134
Black smoke 55, 59–60, 97,
    99
Black smokers 101–03, 121,
    123
    ambient light 134–37
    damage to subs from
       105
    death of 110
    fauna 110–13, 129–35
    fluids 107–08
    height 108–09
    maneuvering *Alvin* at
       103–05, 110–13
    Mid-Atlantic Ridge
       108
    safety 105–07
    sulfide flanges 109–10
    water temperature
       105–07
Blood, tubeworm 81–82
Brisingid sea stars 112–13
Broken Spur 57, 89
    animals 98

    depth at 94–95
    discovery 93
    first dives at 93–98
    sediment 96–97
    Spire, The 58
    Wasp's Nest 58
    water temperature 100
Burke 60
burkensis 60
Byssus 84

caldariensis 62
California 52–53
California Flying Fish
    91–92
Cann, J.R., 130
Carbon dioxide, and tube-
    worms 81–82
Carson, Rachel 9
Cavanaugh, Colleen 80
Certification dives 153
Chace, Fenner A., 128
Challenger Expedition 11
Charlie Gibbs Fracture
    Zone 154
Chemosynthesis 56
Chorocaris 60
Clam Acres 57, 85
Clams 15, 71
    and crabs 85
    giant 57, 85
    graveyard 85
    at Rose Garden 75

Clambake 85–86
Copepods 62–63, 66
Copper sulfides 56
Coral 122–23, 152–53
    on seamount 52
Crabs 13–15, 98
    carapace color 147
    and clams 85
    and copepods 63
    food supply 123–25,
      147
    life stages 147–49
    and tubeworms
      77–79, 116, 147
Crassostrea 10
Crustaceans 57, 62
Curie temperature 118

Dandelions 69–70
Dante 104
Dead Dog 60
Decompression, and animals
    146–48
Deep submergence vehicle
    operator 6
Deep-Water Dumpsite–106,
    160, 164–67
Delaney, John xi, 117–18,
    121, 135–36
Depth of dives, *Alvin* 15,
    29, 32, 94–95
Descent 32, 37–38
Dudley sulfide edifice 60

East of Eden 57
East Pacific Rise
    amphipod 67–69
    Clam Acres 57, 85
    crabs 149
    Melville 14–15
    ocean crust 74
    tubeworms 77
    volcanic eruptions
      116, 123
Ectoparasites 63
Edmond, John xi, 129
Eiffel Tower 58
Embley, Bob 117–18
Endeavour Hydrothermal
    Field 60
Endeavour Segment 104, 135
Enteropneusts 69

Feather dusters 61
Fissures 47
Floc Site 116, 118–19
Flow Site 118
Foster, Dudley xii, 26–27,
    60, 136
fosteri 60
French-American collabora-
    tions 58, 91

Galapagos 69, 156
Galapagos Spreading Center
    57, 75

Index

Galathea Expedition 11
Gamma globulin shot
   159–60
Garden of Eden 57
Gastropods 66
Gay Head–Bermuda Tran-
   sect 3
Geopoetry 73
Georges Bank 153
Gills, mussels 84
Godzilla 108–09
Gore, Robert 13
Grassle, Fred xi, 18, 162,
   165
grasslei 60
Greenland-Iceland Rise
   154
Grieve, Bob 94
Grotto 104
Guaymas Basin 108
   Angel Rock 108
   and copepods 63
   plastic trash bags
      169
Gulf of California 63, 169

H-bomb 30
Half Mile Down 7
Harvard University 80
Hatteras Canyon 152
Hawaii, University of 144
Hemoglobin 82
Hess, Harry 73

Hessler, Robert xi, 60, 162
hessleri 60
Hickey, Pat xii, 27, 94,
   98–100, 136
Hole-to-Hell 59, 123
   volcanic eruption
      116
Holidays at sea 43
Hollis, Ralph xi, 27
hollisi 60
Hot springs 55
Hudson Canyon 152
Humes, Arthur 63, 66
Hydrogen sulfide 56, 81
Hydrothermal springs
   56–57, 74–75
Hydrothermal vents 14
   death of 85
   deepwater circulation
      56
   first discoveries of 57
   light conditions
      135–38
   new species at 57,
      62–63
   sulfides 56–57, 81
   vent fields 57–58, 93
   vent fluids 55–67,
      107–08

Iceland-Faero Rise 154
Iron oxides 97, 110
Iron sulfides 56, 110

177

Jan Mayen 155
Jean Charcot 91
Jericho worm 125
jerichonana 60
Johnson, Paul 117–18
Juan de Fuca Ridge 135
    Axial Seamount 60, 119
    Endeavour Segment 104
    Godzilla 108–09
    ocean crust 74
    Pipe Organ 108
    sulfide flanges 110
    volcanic eruption 115–26

Lamont-Doherty Earth Observatory 93
Langmuir, Charlie 93
Lava 4
    bathtub rings 50
    at Broken Spur 94–96
    color 119
    Curie temperature 118
    Floc Site 118–19
    Flow Site 118
    flows 47–51
    magnetism 118
    and mussels 83, 85

Source Site 118, 121
    volcanic eruptions 116–19
Leeds, University of 130
Light, underwater 32, 133–37, 164
    sunlight, absence of 56, 133
Limpets 60, 65, 144
    and squat lobsters 66
Lobo 104
Lucky Strike 57, 93
Lulu 15–16, 30

Manganese 122
Manipulators 31, 51, 113
    basalt pillows 95
    gathering live tube-worms 78–79
    sulfide spire 99
    video camera 110
Maps 151–52
Martha's Vineyard 3
Martin, Jody 60
Massachusetts Institute of Technology (MIT)/ Woods Hole Oceano-graphic Institution Joint Program 12, 17
Megalopae 147–49
Melville 14
Mexico 15, 71

Mid-Atlantic Ridge 90,
    153–56
    Broken Spur 93
    dominant features of
        134
    major vent sites
        57–58
    shrimp 71, 134
    Snake-Pit 108
Mid-Atlantic Rift 128
Mid-ocean ridges 51,
    74–75
Middle Valley 60
Mohn's Ridge 154
Molgula 10
Moose 108–09
Morse code 34
Mud 160
    collection of
        162–63
    and sewage sludge
        166
Mud dunes 40
Murton, Bramley 93–94,
    96, 99– 100
Museum of Nature 79
Mussels 57, 71, 75
    byssus 84
    lava flow 85
    movements of 84–85
    at Rose Garden 85
    and tubewonns
        83–85
    on whale bones 144

National Geographic 75
National Oceanic and
    Atmosphere Administra-
    tion (NOAA) 119, 128
Nautile 35
Nereid polychaete 63
New Jersey 165–66
New species 60–63

Oasis expedition 60
Oasisia alvinae 60
Ocean crust 73
Oceanographer Canyon 152
Ophiolites 118
Operator's Manual, Alvin 36
Ore 56–57
Overhaul, *Alvin* 22–23, 153
Oxygen, *Alvin* 33, 36
Oxygen, tubeworm 81

Paralvinella 60
Parapodium 61, 64
Passengers, *Alvin* 35–36, 42
Phytoplankton 164, 166
Pilot-in-training (PIT),
    23–28
Pipe Organ 108
Pleopods 148
Polychaete worms 57,
    144–46
    Alvinella 60, 64
    alvinellids 63, 122

Polychaete worms (*Cont.*):
    Broken Spur 97–98
    Nereid polychaete 63
    Paralvinella 60
Polydora 10
Polynoid polychaetes 60
pompejana 62
Ponta Delgada, Azores 93
Pressure specifications 34
Pressure transducers 38
Protozoan 67
Puerto Rico Trench 153
Push cores 94

Rag worm 63
Red shrimp 53
Redheifer, Professor 17
Reykjanes Ridge 154
Reykjavik 155
Rimicaris chacei 128
Rimicaris exoculata 128,
    131, 134–35
Rocks, collected undersea 54
Romanche Gap 156
Rona, Peter 128
Rose Garden 57, 75, 83–85,
    156
    enteropneusts 69
Rutgers University 12, 162

Safety, *Alvin* 37–41, 105–07
San Nicholas Basin 53

Sanders, Howard xi, 162
sandersi 60
Santa Cruz Basin 53
Santa Cruz Island 53
Saracen's Head, The 58–59
Saxon, Charles 160–61
Scripps Institution of
    Oceanography 14, 162
Sea cucumbers 53–54, 166
Sea pens 53
Sea spiders 54–55
Sea urchins 166–67
Seafloor mapping 4
Seafloor markers 123
Seafloor spreading 73–74
Sediment 96–97, 159–67
    age 74
Serpuhd worms 61, 144
Sewage sludge 159–60,
    164–67
Sharks 144
Shellback 156–58
Shipboard refuse 170
Shrimp 127
    Alvinocaris 60
    Atlantic 127–29
    at black smokers
        129–30
    and copepods 63
    edibility 130
    eyes and vision
        131–38
    Mid-Atlantic Ridge
        71, 98–100

new species 128, 131,
134
nutrition 129–30
Pacific 127–29
and water temperature
127–28
Siphonophores 70
Siphonostomes 62–63
Smith, Craig 144–45
Smith, Milt 136
Smithsonian Institution
13–14, 128
Snail, *Alviniconcha* 60
Snake-Pit 57, 93
Beehive 58, 108–09
shrimp 100
Sonar 116, 144
SOSUS 116–17, 122
Sounds, underwater 37
Sound Surveillance System
116
Source Site 118, 121
Southern Cross 156
Spaghetti worms 69–71
Sponges 96, 155
and tubeworms
122–23
Squat lobster 15, 122
and limpets 66
Stars
Brisingid 112–13
brittle 98, 144
Statue of Liberty, The 58
Suction hose 99

Sulfide 57, 79
at black smokers
103
Lucky Strike 93
and tubeworms 81
Sulfide flanges 109–10
fauna 111
Szuts, Ete 132

Terminal velocity 38
Thermal plumes 107
Thermochne 41
Thresher nuclear submarine
30
tibbetsi 60
Tibbetts, Paul 143–44
Titanic 30
Titanium samplers 94
Titanium storage spheres
108
Tivey, Maurice 117–18
Trans-Atlantic Geotraverse
(TAG) 58, 93, 100
Alvin (vent site) 58
Kremlin 58
Mir 58
Transponder 89–91
Trilobite Factor 151, 158
Trophosome 81
Tubeworms 15, 71, 122,
125
bacteria 81–83
and copepods 63

Tubeworms (*Cont.*):
   and crabs 77–79, 116,
      149–50
   features, absence of 80
   gathering live 77–78,
      146
   manganese stained
      122
   and mussels 83–85
   Oasisia alvinae 60, 65
   plume 76, 81–82, 146
   Rose Garden 57, 75,
      82–85
   and sponges 122–23
   sulfur 81
   Tevnia 65
   trophosome 81
   Venture Hydrothermal
      Fields 77
   volcanic eruptions and
      116, 122

U.S. Navy 6, 116
Underwater telephone
   (UQC) 36–37
University of California,
   Los Angeles, 17
Unmanned vehicles 41

vandoverae 60
Venture Hydrothermal
   Fields 77

Video camera, *Alvin* 110
Virgin Mound 60
Volcanic eruptions 101,
   115
   East Pacific Rise
      116
   Juan de Fuca Ridge
      115–16
   submarine mountain
      ranges 117
vulcani 62

Waren, Anders 66
Washington, University of
   109, 117, 136
Water color, undersea
   32
Water temperature 41,
   119–21
   Broken Spur 100
   gradient at black
      smoker 106–07
   hot springs 55,
      127–28
   and light levels 134
   and shrimp 128
   and sub damage
      104–05, 107
   temperature probe 75,
      105–06
Whales 142–44
Williams, Austin 14,
   128

Index

Wireless acoustic link
    15
Woods Hole Oceanographic
    Institution 3, 6, 12
        Alvin overhaul 22
        biologists 162
        copepods 63
        graduate program
            17

Marine Biological
    Laboratory 132
    shrimp edibility test
        130

Zinc sulfides 57
Zooplankton 164
Zoea 149